Advances in Industrial Control

Advances in Industrial Control is a series of monographs and contributed titles focusing on the applications of advanced and novel control methods within applied settings. This series has worldwide distribution to engineers, researchers and libraries.

The series promotes the exchange of information between academia and industry, to which end the books all demonstrate some theoretical aspect of an advanced or new control method and show how it can be applied either in a pilot plant or in some real industrial situation. The books are distinguished by the combination of the type of theory used and the type of application exemplified. Note that "industrial" here has a very broad interpretation; it applies not merely to the processes employed in industrial plants but to systems such as avionics and automotive brakes and drivetrain. This series complements the theoretical and more mathematical approach of Communications and Control Engineering.

Indexed by SCOPUS and Engineering Index.

Proposals for this series, composed of a proposal form (please ask the in-house editor below), a draft Contents, at least two sample chapters and an author cv (with a synopsis of the whole project, if possible) can be submitted to either of the:

Series Editor

Professor **Michael J. Grimble**:
Department of Electronic and Electrical Engineering, Royal College Building, 204 George Street, Glasgow G1 1XW, United Kingdom;
e-mail: m.j.grimble@strath.ac.uk

or the

In-house Editor

Mr. **Oliver Jackson**:
Springer London, 4 Crinan Street, London, N1 9XW, United Kingdom;
e-mail: oliver.jackson@springer.com

Proposals are peer-reviewed.

Publishing Ethics

Researchers should conduct their research from research proposal to publication in line with best practices and codes of conduct of relevant professional bodies and/or national and international regulatory bodies. For more details on individual ethics matters please see: https://www.springer.com/gp/authors-editors/journal-author/journal-author-helpdesk/publishing-ethics/14214

Yang Shi · Chao Shen · Henglai Wei ·
Kunwu Zhang

Advanced Model Predictive Control for Autonomous Marine Vehicles

Springer

Lyapunov-based MPC for the dynamic positioning and trajectory-tracking control design. The optimization problem and a solution methodology are described. An example illustrates the type of results that can be achieved.

Chapter 5 considers the path-following problem of an autonomous underwater vehicle, in which the path following is the main task, and the speed profile is considered a secondary task. A multi-objective MPC approach is followed, that is of interest. The results are illustrated in a simulation that considers the robustness of the solution and this is as important as the performance achieved. The introduction of ocean current disturbances into the problem is considered in Chap. 6. A distributed model predictive formation of the tracking control is proposed.

Chapter 7, on robust distributed MPC platooning control for autonomous surface vehicles, follows on from research on road vehicle systems, and in both cases, energy can be reduced to extend range and safety can be improved. Chapter 8 deals with the important practical problem of implementation of nonlinear MPC algorithms for tracking control. The problem formulation and numerical solution of the optimization problem are discussed, and various designs are compared in simulation. The computing times are compared, and robustness is not forgotten.

Chapter 9 summarizes the main contributions. It also discusses future research directions and includes a look at some fundamental questions, both of which should be useful to research engineers and developers.

The authors have provided a useful introduction to the application problems and suggested very suitable control solutions. This text is in a very important application area that is highly likely to grow substantially. In some application problems, the use of advanced controls may be considered optional, but in areas of autonomous underwater systems the more advanced control methods may be essential if performance goals are to be met. For example, there is the need to limit battery loading and extend the range of autonomous vehicles, and this suggests some form of optimization will be needed. The monograph should be useful to both vehicle control designers and to students interested in the theoretical methods and tools that can be applied.

Glasgow, UK
August 2022

Michael J. Grimble

Preface

The ocean covers about 72% of the earth's surface, but most of the area has not yet been explored. There is an increasing demand for advanced technologies and equipment to explore and exploit the ocean for broad applications, such as tidal energy harnessing, deep ocean exploration, ocean resource investigation, emergency operation support, sustainable ocean environment protection, etc. These challenges and novel applications spur a surge of interest in the research and development of autonomous marine mechatronic systems. Autonomous marine vehicles (AMVs) are those systems with the integration of electronics, mechanics, advanced sensors and actuators, and control algorithms applied in the maritime environment. AMVs play an essential role in performing marine tasks such as offshore inspections, ocean exploration, and transportation, which require effective and reliable motion control systems. Motion control design for AMV systems is challenging, especially when the physical constraints, the control performance, the safety, and the robustness of these systems need to be simultaneously considered. In the past several decades, various advanced control strategies have been proposed for some representative systems in this area. Among these approaches, model predictive control (MPC) is an emerging and promising control scheme.

The essential idea of MPC can be traced back to the 1960s. Until the 1980s, MPC was successfully applied to industrial applications in dealing with constrained multi-variable control problems, especially in process control. Since then, MPC has started to emerge as an advanced control technique and to attract increasing interest concurrently both from academia and industry. MPC has received considerable research attention and achieved successful applications in both practical and theoretical dimensions. Its success is essentially and mainly due to its ability for handling physical constraints, multi-variable interactions, optimal performance, external disturbances, parametric uncertainties, network-induced constraints, cyber attacks, etc., under a unified framework. In general, when applying the MPC framework to mechatronic systems, there are some particular issues and challenges to overcome in the analysis and design. Firstly, we need to comprehensively consider the practical issues, physical constraints, system characteristics, control objectives and tasks, etc. Secondly, we should choose the suitable MPC technique or a seamless integration of several MPC techniques for fulfilling the specific task. Thirdly, we need to rigorously analyze the theoretical properties of the resulting closed-loop system by applying the MPC strategy. Finally, we should consider the practical implementation of the MPC strategy in real-time control systems. It is underscored that the MPC studies are not merely driven by theoretical curiosity, but also are tantamount to enhanced performance with practical value.

There is also increasing awareness from the AMVs community of the need for accommodating the physical (states and control input) constraints and achieving certain optimal performance. This sets the stage for the emergence of a new application-oriented research—MPC for the motion control of AMVs—aiming at supplying the MPC algorithm design, analysis of the theoretical properties, and the implementation guidance. Our intention through this book is to provide a timely account as well as an introductory exposure to some of our main developments in the application of MPC for AMVs. The book will present some design and analysis of MPC-based motion control algorithms for the nonlinear AMV systems for different marine missions. We also hope that the book may further spark new ideas, thoughts, and directions on the subject.

The book consists of nine chapters. Chapter 1 provides a comprehensive overview of the motion control of AMVs and a brief introduction to MPC. Chapter 2 reviews the modeling of the AMV system and explores some important properties associated with the system model. Chapter 3 studies the combined path planning and trajectory-tracking control of an autonomous underwater vehicle (AUV). A unified receding horizon optimization framework is developed with a novel spline-based path planning method and the nonlinear model predictive tracking controller design. Chapter 4 develops a Lyapunov-based MPC approach for the dynamic positioning and trajectory-tracking control of the AUV. Chapter 5 solves the path-following problem of the AUV by developing a multi-objective MPC framework. Chapter 6 focuses on the formation-tracking control problem of multi-AUV systems with external disturbances. A distributed Lyapunov-based MPC algorithm is developed to solve this problem. Chapter 7 applies a robust distributed model predictive platooning control approach for a group of heterogeneous autonomous surface vehicles (ASVs) with input constraints and bounded external disturbances. Chapter 8 investigates how to reduce the heavy computational burden and increase the speed of solving the formulated optimization problem, aiming to pave the way toward the practical application of MPC to AMVs in real time. Two novel numerically efficient algorithms, namely the modified C/GMRES (continuation/ generalized minimal residual method) algorithm and the distributed NMPC algorithm, are presented to reduce the computational complexity. Chapter 9 summarizes the work in this book and provides some potential future research directions. Moreover, for it to be self-contained, a detailed description of the software implementation and simulations is provided in the appendix.

Acknowledgements

The authors would like to thank all those who have helped in accomplishing this book.

The writing of this book as well as our research on the application of MPC for AMVs has benefited greatly from interactions and collaborations with many colleagues. We wish to express our heartfelt gratitude to Demin Xu, Weisheng Yan, Mingyong Liu, Huiping Li, Jian Gao, Changxin Liu, Brad Buckham, and Colin Bradley, with whom we have enjoyed pleasant and fruitful collaborations. In 1993, the first author started his graduate studies under the supervision of Prof. Demin Xu in the School of Marine Science and Technology at Northwestern Polytechnical University (Xi'an, China). The second and the third authors also graduated from the same school and then joined the Applied Control and Information Processing Laboratory (ACIPL) at the University of Victoria to pursue their Ph.D. degrees. Three of them together dedicate the book to Prof. Xu as a way of expressing their deep gratitude and indebtedness, for guiding them into the amazing world of marine engineering and control engineering. Support from Natural Sciences and Engineering Research Council (NSERC) has been very helpful and is greatly acknowledged. Finally, all the authors would like to greatly express special thanks to their family members.

Victoria, Canada — Yang Shi
Ottawa, Canada — Chao Shen
Victoria, Canada — Henglai Wei
Victoria, Canada — Kunwu Zhang

MIMO	Multiple-input multiple-output
MOMPC	Multi-objective model predictive control
MPC	Model predictive control
MSE	Mean square error
NED	North–East–Down
NLP	Nonlinear programming
NMPC	Nonlinear model predictive control
NN	Neural network
OCP	Optimal control problem
PD	Proportional–derivative
PF	Path following
PID	Proportional–integral–derivative
PMP	Pontryagin's minimum principle
PO	Pareto optimal
QP	Quadratic programming
RHO	Receding horizon optimization
ROA	Region of attraction
ROV	Remotely operated vehicle
SQP	Sequential quadratic programming
TA	Thrust allocation
TRD	Trust-region-dogleg
UAV	Unmanned aerial vehicle
UUB	Uniform ultimate bounded
UUV	Unmanned underwater vehicle
WS	Weighted sum
WS-MOMPC	Weighted-sum-based multi-objective model predictive control
ZPE	Zero-path-error

Chapter 1
Introduction

1.1 Autonomous Marine Vehicles

1.1.1 Overview

Oceans cover two-thirds of the Earth surface and have a huge impact on our ecosystem. Traditionally, they act as the source of food, provide warmth and natural resources, and sustain the ocean ecosystem by maintaining biodiversity. With the development of ocean science, their ecological, economic, and social importance are now better understood. On the other hand, ocean activities are closely related to some deadly natural phenomenons such as tsunami, earthquake, and hurricane. Hence, the constant monitoring of the ocean state becomes an urgent necessity and will definitely benefit mankind in terms of minimizing the loss due to natural disasters, maximizing the harvest from the oceans, and more.

Autonomous marine vehicles (AMVs) generally include autonomous surface vehicles (ASVs) and unmanned underwater vehicles (UUVs) as shown in Figs. 1.1 and 1.2; these advanced tools enable the ocean monitoring to go far beneath the ocean surface, collect diverse first-hand data, and see how the oceans behave (Fossen 2002). Clearly, the manned marine technology was firstly focused. Since 1962 when the first submarine was constructed (Roberts and Sutton 2006), dramatic progress has been made in the design and manufacturing of manned marine vehicles. However, the intrinsic weakness of reliance on human pilots limits its applications. In contrast, advances in navigation, control, computer, sensor, and communication technologies have turned the idea of AMV into reality.

In particular, the UUVs can be categorized into two groups:

- Remotely operated vehicles (ROVs) are tethered underwater robotic vehicles. ROVs require instructions from human operators who locate in the support vessel during the execution of tasks. An umbilical cable is therefore needed to carry power, relay control signals, and transmit sensor data.

Y. Shi et al., *Advanced Model Predictive Control for Autonomous Marine Vehicles*, Advances in Industrial Control, https://doi.org/10.1007/978-3-031-19354-5_1

Fig. 1.1 Autonomous surface vehicle—the Otter ASV. Photo courtesy of Maritime Robotics. ©2022 Maritime Robotics, reproduced with permission from Maritime Robotics (2022)

- Autonomous underwater vehicles (AUVs) are tether-free underwater robotic vehicles. AUVs are powered by onboard batteries or fuel cells, equipped with navigation sensors, and execute preprogrammed missions without being constantly supervised or controlled by humans.

In recent years, UUVs have achieved great success in many ocean-related scientific fields such as marine geoscience (Wynn et al. 2014; Yoerger et al. 2007; Yoon and Qiao 2010; Bachmayer et al. 1998), offshore industry (Wernli 2000; Naeem et al. 2004; Elvander and Hawkes 2012) and deep-sea archeology (Bingham et al. 2010; Roman and Mather 2010). Compared to ROVs, AUVs have higher level of autonomy and demonstrate the following strengths:

- AUVs have the much wider reachable scope. AUVs are more mobile platforms and can execute oceanic missions that need to travel a long distance, e.g., polar region survey beneath the ice sheet, or that need to be performed in dangerous areas, e.g., submarine volcanism data acquisition.
- AUVs avoid many technical issues related to the tether cable. The chaotic drag force induced by the cable makes the vehicle difficult to control. The drag force will become unmanageable as the tether length increases. Moreover, the communication latency greatly influences the control of the vehicle. In contrast, the absence the umbilical cable enables the real-time control of AUVs.
- AUVs reduce the operational cost. Unlike ROVs which need human operators to perform the task, AUVs require the minimum amount of human intervention. Therefore, it is likely to cut out a large portion of the operational cost by reducing the number of human staff in a task.

Fig. 1.2 Unmanned underwater vehicle—BlueROV2. Photo courtesy of Blue Robotics. ©2022 Blue Robotics, reproduced with permission from BlueRobotics (2022)

On the other hand, the absence of tether cable brings challenges in the power supply, underwater navigation, and automatic control aspects. With the developments of new technologies in these areas, AUVs have extensive application prospect. There are two configurations for the shape design of AUVs:

- The conventional slender body AUVs have efficient hydrodynamic properties and are best suited for oceanic missions that need to travel with a high speed or a long distance. These conventional AUVs are usually equipped with propellers to drive in the direction of the principal axis and with control surfaces (rudders) to perform maneuvers. Therefore, they have lower number of control variables than the motion degrees of freedom (DOF), i.e., underactuated. Hence, they are easy to control only along straight lines. Examples of AUVs with this configuration include REMUS (Allen et al. 1997) and ODYSSEY (Desset et al. 2005).
- The design of open-frame AUVs was recently borrowed from ROVs in order to enable the omni-directional mobility. This configuration usually contains redundant thrusters to provide more control DOF than the motion, hence the AUV can perform good low-speed maneuvers in cluttered environments. With the omni-directional mobility and the possible artificial intelligence, AUVs are potentially capable of performing complicated jobs. Examples of AUVs with this configuration include Ictineu (Ribas et al. 2010) and Smart-E (Meyer et al. 2013).

The control system can be viewed as the brain of the AMV and is responsible for the autonomy of the vehicle. The term *control* here has a broad sense, including but not limited to

- Motion Control: The low-level system control, focusing on the input/output of the vehicle motion and the closed-loop properties such as stability and robustness.
- Mission Control: The high-level behavioral control pipeline which is elaborately designed to accomplish the predefined missions or tasks in a fully autonomous way. Each step will be triggered by sensor measurements.

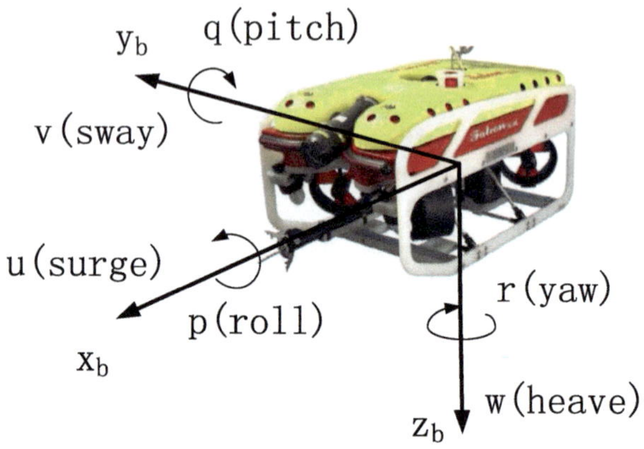

Fig. 1.3 Motion variables of an AMV. Photo courtesy of Saab Seaeye Ltd. © 2022 Saab Seaeye Ltd., reproduced with permission from Saab Seaeye Ltd. (2022)

- Power Management: The control that aims at optimally distributing the onboard power with various system and mission constraints. For the AMVs with onboard solar power systems (Jalbert et al. 2003), it may also integrate the recharging plan into the power management.
- Cooperative Control: The control that emphasizes the coordinated behaviors among a group of AMVs, e.g., formation-tracking control and platooning control, in order to accomplish the tasks in an optimal way.

Among all above control categories, it may be hard to distinguish one specific type to be more important than the others. However, there is little controversy over the following statement: *Motion control is the most fundamental research study for the control of AMVs*. High-level mission control tasks and cooperative control tasks could only be achieved through the motion control of each individual vehicle.

1.1.2 The Motion Control Problems

The motion of an AMV in the three dimensional workspace can be described in six DOFs. The six independent variables which uniquely determine the motion of the vehicle are known as 'surge', 'sway', 'heave', 'roll', 'pitch', and 'yaw' (see Fig. 1.3). Note that the ASV is generally described in 3-DOF, i.e., 'surge', 'sway', and 'yaw'.

The motion control of AMVs aims to regulate the motion variables to their desired values (called set-points) which are determined by the high-level motion planning system. According to different types of set-points, the motion control problems can be classified into three basic categories:

- When the set-points are time-invariant, it belongs to the point stabilization category. In context of AMV motion control, the heading control, depth control and dynamic positioning (DP) control belong to this category. Specifically, the DP refers to the automatic control of AMV to reach and then maintain its position and orientation at a fixed point, probably with significant external disturbances from winds, waves and ocean currents.

- When the set-points are time-varying, it belongs to the trajectory-tracking control category. The trajectory-tracking controller is designed to steer the AMV position and velocity to converge and then precisely track the desired trajectory (time-varying set-points) that is obtained from the high-level motion planner.
- When the set-points are time independent but describe the geometric relationship among them, it is a path-following (PF) problem. The PF control refers to the automatic control that moves the AMV along a specified path in the workspace, but there is no requirement on when the AMV should be where.

Practically, the AMV motion control problems can be solved at two levels:

- The velocity-level solution. The velocities of the AMV are the control signals. The motion control system determines the desired linear and angular velocities that steer the AMV to achieve the control objectives. The control of thrusters that generate the desired velocities is assumed to be solved readily. At this level, only the kinematics of the AMV motion need to be considered.
- The force-level solution. The forces and moments that cause the AMV motion are regarded as the control inputs. The motion control system determines the desired propulsive force for each thruster according to the AMV state and control objectives. At this level, both kinematics and dynamics of the AMV motion need to be considered.

The two levels of solutions reflect the design trade-off between precision and complexity. Since the AMV dynamics are not considered, the control algorithm design for the velocity-level solution can be simplified significantly. For the same reason, however, the control precision in velocity-level solutions may be compromised, especially for the trajectory tracking and PF applications. In contrast, the force-level solutions are expected to have high precision. But since it considers the complex dynamics of AMV motion, a modeling and system identification is required before controlling the vehicle. Sometimes, the identified AMV dynamic model may be highly nonlinear and time-varying, which makes the control system design very challenging or even intractable.

1.2 Model Predictive Control

Optimal control is an important research direction in control engineering and applied mathematics. Early theoretical results mainly include Bellman's principle of optimality (Bellman 1957), Pontryagin's minimum principle (PMP) (Kalman 1960; Arthur and Ho 2018), and linear quadratic regulator (LQR) (Kalman 1960). However, these control methods cannot handle system constraints on the system state and/or on the control variables, which evokes the keen interest in studying MPC. On the other hand, in many industries such as petrochemical industry, the requirement of optimal process control to chase a maximum profit stimulates the growth of MPC. The optimal operation is often near or on the boundary of the permitted operation region, and the system constraints have to be considered explicitly.

1.2.1 The Receding Horizon Control Strategy

Generally speaking, the MPC is a control strategy which determines the control action by recursively solving finite horizon optimal control problems (OCPs) and takes into account all the system constraints during the control (Mayne et al. 2000; Mayne 2014). Consider a general nonlinear system

$$\dot{x}(t) = f(x(t), u(t)), \quad x(0) = x_0, \tag{1.1}$$

where $f : \mathbb{R}^n \times \mathbb{R}^m \to \mathbb{R}^n$ represents the underlying nonlinear dynamics. Without loss of generality, the origin is assumed to be the equilibrium of interest, and the control objective is to steer the system state to the origin.

The finite horizon OCP is defined as follows:

$$\begin{aligned} \min_u \ J(x, u) &= \int_0^T \ell(x(s), u(s))\mathrm{d}s + g(x(T)) \\ \text{s.t.}\ \dot{x}(s) &= f(x(s), u(s)), \\ x(0) &= x_0, \\ x(s) &\in \mathbb{X}, \\ u(s) &\in \mathbb{U}, \end{aligned} \tag{1.2}$$

where $J(x, u)$ is the cost function consisting of the stage cost $\ell(x, u)$ and the terminal cost $g(x)$; T is the prediction horizon and x_0 is the current measured system state; $\mathbb{X}$ and $\mathbb{U}$ are compact sets, representing the system constraints on the state and on the control, respectively.

The MPC is implemented in a receding horizon control paradigm which can be briefly described as follows:

- ☐ At the sampling time instant, the OCP (1.2) is solved, which obtains the solution curve $u^*(s)$, $s \in [0, T]$.
- ☐ The first portion of the solution curve, $u^*(s)$ for $s \in [0, \Delta t]$, is actually implemented to control the nonlinear system, where Δt is the sampling period.
- ☐ At the next sampling instant, the OCP (1.2) will be solved again with the system state measured and used as the new initial condition.

Since MPC is implemented on digital computers, the OCPs need to be discretized and then solved by iterative numerical algorithms.

1.2.2 *Stability of MPC*

The MPC is implemented by recursively solving finite horizon OCPs. While the finite time horizon makes the solving of OCPs numerically tractable, it throws the closed-loop stability into question. Optimality does not necessarily lead to stability. As shown in (Arthur and Ho 2018), even for linear systems with no constraints, the finite horizon LQR can be destabilizing. Similar situation occurs in MPC as well.

1.2.2.1 A Brief Review of Stability Results in General MPC Theory

Early days, however, the closed-loop stability was obtained in most process control applications, certainly after tuning. This is because the prediction horizon was normally sufficiently long. Primbs and Nevistic (1997) showed that for linear systems, the length of a stabilizing prediction horizon can be precomputed with or without the state and control constraints. For nonlinear constrained systems, Alamir and Bornard (1995) established the asymptotic stability for a given region of attraction (ROA) with a sufficiently long prediction horizon and a shorter control horizon.

On the other hand, the computational complexity of the OCP increases exponentially as the prediction horizon increases. In many applications such as AMV motion control, it is not affordable to use a too long prediction horizon. Since the closed-loop stability is an overriding requirement, during the past forty years, control theorists and practitioners have been devoting a significant amount of effort to the development of stable MPC schemes that do not rely on long prediction horizons. Among the existing results, seeking stabilizing conditions using the terminal cost, terminal constraints, and an associated terminal controller (known as the terminal triple) is the most popular approach (Mayne et al. 2000). Terminal equality constraint was firstly imposed. Keerthi et al. demonstrated in (Keerthi and Gilbert 1988) that the optimal value function of the infinite horizon OCP can be approached by that of a finite receding horizon approximation with the terminal state constraint $x(T) = 0$. Imposing the terminal equality constraint, however, adversely affects the feasibility of the OCP, and in turn requires a long prediction horizon, which is undesired. Therefore, the dual-mode MPC scheme (Chisci et al. 1996; Scokaert et al. 1999) were proposed. In the dual-mode MPC scheme, a terminal set constraint $x(T) \in X_{\mathrm{f}}$ is used in place of the former equality constraint. Once the system state x enters the terminal set X_{f}, a local feedback controller $\kappa_{\mathrm{f}}(x)$ will take over the stabilization task and steer the system state to the origin. However, the implementation of the dual-mode MPC is complicated, and it also loses some degree of optimality. Therefore, in the most recent MPC proposals, the terminal triple is used together to establish the stability condition. In an excellent MPC review paper, Mayne et al. (2000) summarized the stabilizing conditions in most MPC proposals and distilled the widely accepted principle of stability in terms of four mild assumptions. Once the four assumptions are satisfied, the optimal value function of the OCP can be shown a valid Lyapunov function for the closed-loop nonlinear system, hence guaranteeing the stability.

In the new century, with the significant development of stabilizing conditions without imposing a terminal constraint (Mayne 2014), there emerges an interesting discussion in the MPC community: On the one hand, the conservativeness of MPC can be fairly relaxed (Grüne 2012) if there is no terminal constraints; on the other hand, the necessity of a terminal constraint is emphasized in (Mayne 2013) for guaranteeing the recursive feasibility. Other MPC schemes which do not exploit the terminal triple mainly include variable horizon MPC (Michalska 1997), contractive MPC (Liu et al. 2014), and Lyapunov-based MPC (LMPC) (de la Peña et al. 2008).

1.2.2.2 Stability Results in MPC for AUV Motion Control

While a number of MPC solutions have been proposed for AUV motion control problems (e.g., Oh and Sun 2010; Pavlov et al. 2009; Veksler et al. 2016; Naeem et al. 2004), none of them include the stability analysis. This is partly due to the complicated nonlinear dynamic model of the AUV, and partly due to the fact that the MPC theoretical papers focus exclusively on the point stabilization problems while for AUV motion control we may be more interested in the trajectory tracking and PF problems.

In our published work (Shen et al. 2017), the stability of an MPC-based trajectory-tracking controller is explicitly discussed for the first time. By defining an appropriate reference system, we converted the trajectory-tracking control problem to the stabilization problem of the error dynamics and then followed the principle of MPC stability (Mayne et al. 2000) to derive the sufficient conditions for guaranteeing the closed-loop stability. In (Shen et al. 2017), we took a LMPC strategy to deal with AUV trajectory-tracking control. The closed-loop stability is guaranteed by imposing a contraction constraint derived from a Lyapunov-based control law. The design of a stable MPC controller for the AUV DP problem is studied in (Shen et al. 2017), and the MPC solution to the AUV PF control problem is discussed in (Shen et al. 2018).

1.2.3 MPC for the Motion Control of AMVs

A variety of advanced control strategies have been proposed for some representative systems in this area, including AMVs (Shi et al. 2017). Among these approaches, MPC is an emerging control scheme, and it has the advantage to effectively cope with various physical constraints while achieving optimal control performance. In this subsection, we summarize and analyze some recent advances in marine control systems from the perspective of MPC. As shown in Table 1.1, different MPC methods have been developed for AMVs capable of performing a variety of control objectives, including DP (regulation), PF, trajectory tracking, and cooperative control.

Table 1.1 Representative works on the motion control of AMVs using MPC

Applications	Representative works
Dynamic positioning	Wahl and Gilles (1998), Perez et al. (2000), Li and Sun (2011), Veksler et al. (2016), Fernández et al. (2016), Tsopelakos and Papadopoulos (2016), Li and Yan (2016), Shen et al. (2017), Nascimento et al. (2009), Zheng et al. (2020)
Path following	Li et al. (2009), Pavlov et al. (2009), Oh and Sun (2010), Alessandretti et al. (2013), Kapetanović et al. (2017), Zhang et al. (2017), Shen et al. (2018)
Trajectory tracking	Yan and Wang (2012), Guerreiro et al. (2014), Shen et al. (2016), Shen et al. (2016, 2017), Gan et al. (2018), Zhang et al. (2019), Heshmati-Alamdari et al. (2020, 2019), Liang et al. (2020), Shen and Shi (2020)
Path planning	Huynh et al. (2015)
Cooperative control	Zheng et al. (2016), Li et al. (2016), Zheng et al. (2017), Chen et al. (2018), Ferranti et al. (2018), Hung et al. (2020), Wei et al. (2021), Chen et al. (2019), Yang et al. (2021)

- Dynamic positioning: The AMV is required to maintain the predetermined position and heading autonomously with installed thrusters. This basic control task is still challenging due to the nonlinear dynamics and unknown disturbances induced by the environment, such as winds, waves, and currents.
- Path following: The AMV is required to follow a feasible predefined path with desired speed and orientation.
- Trajectory tracking: The AMV is driven to track a spatial and temporal trajectory with a specific time requirement. This problem still needs investigation for underactuated AMVs.
- Cooperative control: A group of AMVs is expected to cooperatively finish a specific control task, such as the formation tracking.

According to different tasks, the applications of MPC for AMV systems have been widely studied in the literature. Some of the representative results are reviewed in the following:

Dynamic Positioning. The DP control is one of the typical control tasks for marine applications [see the survey paper (Sørensen 2011), and the references therein]. Early work on MPC for marine vehicles mainly studies the angle stabilization problem and adopts a one-DoF dynamical system model (Wahl and Gilles 1998; Perez et al. 2000). The control objective is to steer the heading to the desired angle and reduce the roll angle under different external environments. A novel disturbance compensating MPC method is developed to handle the constraint, the disturbances, and the feasibility issues for ASV heading control problem in (Li and Sun 2011). The DP using MPC is studied in (Veksler et al. 2016), considering the dynamic thruster capability, the reduction of energy consumption, and the improved control performance. In order to avoid the complex design of the terminal set (Shen et al. 2017) study DP of AUVs and impose a Lyapunov-based constraint into the MPC optimization problem to guarantee the closed-loop stability. Recent results on robust DP using robust MPC are studied for ASVs with model uncertainties and external environmental disturbances in (do Nascimento et al. 2009; Zheng et al. 2020).

Path Following. Various control methods have been studied for the PF problem of AMVs (Shi et al. 2017). However, the control performance may degrade due to ubiquitously existing constraints on states and inputs. In (Li et al. 2009), the PF problem of marine vessels with the rudder saturation and rate limits has been solved by using MPC. Furthermore, (Oh and Sun 2010) propose a novel MPC method to simultaneously design the guidance and control systems for USVs. A multi-objective MPC method is proposed for AUV to improve the PF performance by considering the flexibility of adjusting the velocity of the vehicle. On the other hand, MPC has been adopted for the fully-actuated marine vehicles (Li et al. 2009; Shen et al. 2018) and underactuated marine vehicles (Pavlov et al. 2009; Oh and Sun 2010; Li and Yan 2016; Zhang et al. 2017; Kapetanović et al. 2017). The issues of environmental disturbances are tackled by employing robust MPC in (Zhang et al. 2017).

Trajectory Tracking. Another typical task of AMVs is trajectory-tracking control, and there exist a number of research results (Shi et al. 2017; Liu et al. 2016). The unpredictable and complex sea environment, highly nonlinear system dynamics, and various constraints give rise to challenges for the tracking controller design. MPC can be applied to solve the challenging trajectory-tracking control problem of AMVs under different scenarios. In (Yan and Wang 2012), a neurodynamic optimization-based MPC method is proposed for the tracking control of underactuated vessels, in which a single-layer recurrent neural network is used to solve the formulated quadratic programming problem. A nonlinear MPC (NMPC) method is presented to address the trajectory-tracking problem of ASVs considering the presence of ocean currents in (Guerreiro et al. 2014). To implement MPC in real time, a modified C/GMRES algorithm has been developed to solve the NMPC optimization problem (Shen et al. 2016).

Shen et al. (2020) further decompose the original optimization problem into smaller size subproblems such that they can be solved in a distributed manner, thus resulting in significantly improved computational efficiency. An integrated MPC scheme of jointly solving the path planning and trajectory-tracking control problem for AUVs is proposed in (Shen et al. 2016). The conventional MPC scheme employs the terminal cost and terminal set to ensure the closed-loop stability, which is hard to design for nonlinear AUVs. Hence, a Lyapunov-based stability constraint is designed and imposed on the MPC tracking problem. The robust MPC method is studied for AMVs subject to external disturbances in the literature (Zhang et al. 2019; Heshmati-Alamdari et al. 2019, 2020). (Liang et al. 2020) investigate the MPC-based trajectory tracking of underactuated AMVs where an auxiliary time-varying tracking controller is employed for the terminal constraint design. The resultant NMPC scheme is validated on a real ASV platform.

Cooperative Control. In order to complete the complex mission, reduce the operational cost, and improve the control performance, distributed MPC (DMPC) for a team of AMVs has become an important research topic in the marine robotics and control field. In contrast with the centralized MPC, DMPC has a lower computational requirement and is suitable for physically distributed multiple AMVs. Hence, DMPC has been used to handle challenging issues, including computational inefficiency, disturbances and uncertainties, imperfect communication, collision avoidance, connectivity maintenance, and heterogeneous cooperation. In (Zheng et al. 2016), a distributed predictive PF controller is proposed for multiple waterborne automated guided vessels. The alternating direction method of multiplier (ADMM) is used to accelerate the convergence rate of the DMPC optimization problem. Furthermore, this method is extended as robust DMPC to consider the external disturbances explicitly. The authors apply the DMPC method to solve the formation-tracking control problem of multiple underactuated AUVs (Li et al. 2016). A distributed Lyapunov-based MPC (DLMPC) method is designed to achieve the formation tracking of multiple AUVs with environmental disturbances while ensuring the collision avoidance (Wei et al. 2021). Due to the limited communication bandwidth among AUVs, an offline quantization design and an online DMPC algorithm with quantization are proposed to solve this issue. The communication burden is alleviated by using the event-triggered DMPC method in (Hung et al. 2020). The 'vessel train' and 'ASV platoon', a special cooperative formation of marine vessels, are developed with the aim of improving the efficiency of waterborne transport in (Chen et al. 2018; Wei et al. 2021).

1.3 Motivations of the Book

Although a significant amount of effort has been devoted to the study of AMV motion control, due to the complexity of the control problem itself, there still exist many unsolved issues in this area. One prominent issue is that the practical constraints on the real AMV system such as limited perceiving, computing, and actuating capabilities are seldom considered in the controller design, because of the inherent limitations of those conventional control methods.

The MPC presents a powerful framework for solving a broad spectrum of control engineering problems and has the capability of handling practical constraints in a systematic manner. However, due to its heavy computational burden, for many years, MPC had been applied to process control problems with slow dynamics only. With the development of computer technology and optimization theory, now the computational barriers have been largely removed. The successful real-time implementation of MPC for UAVs (Baca et al. 2016) and mobile robots (Howard et al. 2010) strongly suggests that the MPC can be applied to AMVs to address the practical constraint issue. Besides, the MPC demonstrates good inherent robustness against model uncertainties and external disturbances, which makes it a perfect solution to AMV control problems. On the other hand, due to the implicit nature of the optimization procedure, the characterization of closed-loop stability for the MPC-based AMV control is challenging and complicated. So far, few control theoretical results on the MPC-based AMV motion control have been reported in the literature. To fill this gap and push forward the application of the advanced model predictive control technology to AMV control systems, this book focuses on the study of the MPC-based AMV motion control problems and attempts to lay the theoretical foundation for the application of MPC to the marine vehicle systems. The main contributions of this book are summarized as follows:

- **Design of nonlinear model predictive control for the integrated path planning and tracking control.** A unified receding horizon optimization (RHO) framework is proposed for solving the integrated path planning and tracking control problem of an AUV. The RHO framework consists of a spline-based path planning module and an NMPC tracking controller. The path planning is formulated into RHO problems which accommodates the practically finite perceiving capability of the AUV. Once the reference path is planned, with a predetermined timing law, it is augmented in order to provide the reference trajectory for each state of the AUV. Then an NMPC tracking controller is designed for the vehicle to precisely track the reference trajectory. Sufficient conditions for closed-loop stability are derived. Finally, an implementation algorithm which seamlessly integrates the path planning and the NMPC tracking control is proposed. With the implementation algorithm, the obtained closed-loop stability of the NMPC tracking control can be preserved. To the best of our knowledge, it is the first time to explicitly conduct the stabilizing conditions of the MPC-based trajectory-tracking control for marine vehicles.
- **Design of Lyapunov-based model predictive control (LMPC) for the dynamic positioning and trajectory-tracking control.** Firstly, an LMPC-based DP control algorithm is proposed for an AUV. A nonlinear PD control law is exploited to construct the contraction constraint in optimization problem that is associated with the LMPC. A quasi-global closed-loop stability property can be claimed for the closed-loop LMPC-based DP control system. Secondly, the LMPC is applied to solve the AUV trajectory-tracking control problem. An auxiliary nonlinear tracking control law is designed using the backstepping technique and then used to construct the contraction constraint. Conditions for recursive feasibility and closed-loop stability are derived. In both DP and tracking control, the thrust allocation (TA) subproblem is solved simultaneously with the LMPC control, which reduces the conservativeness brought by conventional TA solutions. Essentially, the proposed LMPC method builds on the existing AUV control system and incorporates online optimization to improve the control performance. Since the closed-loop stability does not rely on the exact solution of the optimization, the LMPC creates a trade-off between computational complexity and control performance. We can easily control the computational complexity by specifying the maximum iteration number meanwhile guarantee the control performance no worse than the existing AUV motion controller.
- **Design of multi-objective model predictive control (MOMPC) for the path-following control.** A novel MOMPC method is proposed to solve the PF control problem of an AUV. Two performance indexes which reflect the path convergence

requirement and the speed assignment are designed. Then the PF problem can be formulated into the MOMPC framework with the two performance indexes as the objective function. Since the path convergence is usually more important than the speed assignment, two methods which handle objective prioritization are proposed to solve the associated vector-valued optimization problem. The internal relationship between the two methods are explored and the conditions for closed-loop stability are provided. The proposed MOMPC method not only provides a novel scheme to solve the AUV PF control problem, but also lays a foundation for the study of AUV motion control problems with multiple control objectives.

- **Design of distributed Lyapunov-based model predictive control for multi-vehicle cooperative control.** The LMPC strategy is extended to solve the cooperative formation-tracking control problem of a multiple AUV team. In order to limit the computation and communication cost, we develop the LMPC algorithm in the distributed control paradigm. Each and every AUV communicates with its neighbors and optimizes its own behavior based on the information from them. Collision avoidance and explicit disturbance handling are particularly addressed and then seamlessly integrated into the DLMPC algorithm. Recursive feasibility and closed-loop stability under the proposed DLMPC algorithm can be rigorously proved. The proposed DLMPC is optimal, robust, and fully scalable, thus it has shed light on how MPC or general optimizations can be exploited (with theoretical guarantees) in the control system design for multi-vehicle systems.
- **Design of robust distributed model predictive control for the heterogeneous surface vehicle platoon.** A robust distributed model predictive platooning control approach is applied for a group of heterogeneous ASVs with input constraints and bounded external disturbances. The control input for each ASV is composed of two parts: the optimal nominal control input and the ancillary control input. The optimal nominal control input is generated by solving a DMPC problem based on the state information of itself and its neighbors. The offline ancillary control law ensures that the actual system state trajectory evolves in a hyper-tube centered along the optimal nominal state trajectory. A coupled intervehicle safety constraint is designed in the DMPC optimization problem to guarantee inter-ASV collision avoidance. Theoretical results on the feasibility of the proposed robust DMPC algorithm are provided, and the closed-loop systems are proved to be input-to-state stable (ISS). Numerical simulations are performed to illustrate the theoretical results.
- **Design of efficient implementation algorithms for the NMPC-based AUV control.** Two distinct fast implementation strategies are proposed for the NMPC-based trajectory-tracking control of an AUV. The first strategy is based on the numerical continuation method. Assuming that the solution of the associated optimization problem is not obtained at singular points, the NMPC control signals can be approximated without undergoing the successive linearization step which is inevitable in off-the-shelf NLP algorithms, therefore, the computational complexity can be significantly reduced. The convergence of the solution is proved. The second strategy exploits the dynamic properties of the AUV motion and solves the optimization problems in a distributed fashion. The recursive feasi-

bility and closed-loop stability of the distributed implementation are proved. The proposed fast implementation strategies considerably alleviate the heavy computational burden hence greatly increases the possibility of implementing NMPC-based motion control on various AUVs including those with limited onboard computing resources.

1.4 Outline of the Book

This section provides a map of the book showing the readers where and how it validates the claims we have previously made.

Chapter 1 contains the fundamentals and literature reviews of the closely related research fields. It also presents the research background, motivations, and main contributions of this book.
Chapter 2 develops the mathematical model of AUV that will be used throughout the book. Several important properties associated with the developed model are also explored in this chapter.
Chapter 3 studies the combined path planning and tracking control of an AUV. A unified RHO framework is developed with a novel spline-based path planning method and the nonlinear model predictive tracking controller design.
Chapter 4 presents a LMPC framework for the motion control of an AUV. The LMPC controller designs for DP and trajectory tracking are detailed.
Chapter 5 considers the PF control problem of an AUV. A novel MOMPC framework is proposed to handle the objective prioritization.
Chapter 6 extends the MPC results to multi-AUV control problems. The proposed DLMPC method is applied to a multi-AUV system. The formation-tracking control design is detailed with explicit compensation for external disturbances.
Chapter 7 develops robust and safe DMPC algorithm for cooperative heterogeneous ASVs in the presence of uncertainties.
Chapter 8 aims to solve the bottleneck problem of implementing MPC on AUVs: The heavy computational burden. Two novel numerically efficient algorithms, namely the modified C/GMRES algorithm and the distributed NMPC algorithm are introduced to reduce the computational complexity.
Chapter 9 summarizes the work in this book, discusses the opportunities and challenges in MPC and AMV control system design, and provides some potential future research directions.

1.5 Notes and Summary

In this chapter, the AMV control design problems are discussed. The open problems and challenges in this research field are revealed after doing a comprehensive literature review. The motivation of the book is provided with a brief introduction of general MPC theory and its application in autonomous marine control system design. Following this, the contributions and structure of the book are presented. The MPC-based controller design is an emerging research field for marine vehicle systems, and there are many opportunities and challenging problems awaiting to be studied. We picked out several important survey papers to facilitate interested readers who want to jump into this field. For extensive overviews of AMV control designs, the readers are referred to (Fossen 2011; Shi et al. 2017; Yuh 2000) and the references therein. For overview of general MPC algorithms, theoretical developments and useful design methodologies (Mayne et al. 2000; Mayne 2014; Rawlings et al. 2017) can be helpful. For MPC architecture design and industrial applications (Qin and Badgwell 2000; Scattolini 2009) are recommended for further readings.

References

Alamir M, Bornard G (1995) Stability of a truncated infinite constrained receding horizon scheme: the general discrete nonlinear case. Automatica 31(9):1353–1356

Alessandretti A, Aguiar AP, Jones CN (2013) Trajectory-tracking and path-following controllers for constrained underactuated vehicles using model predictive control. In: Proceedings of the 2013 European control conference (ECC), Zurich, Switzerland, 2013. IEEE, pp 1371–1376

Allen B, Stokey R, Austin T, Forrester N, Goldsborough R, Purcell M, von Alt C (1997) REMUS: a small, low cost AUV; system description, field trials and performance results. In: Oceans' 97. MTS/IEEE conference proceedings, vol 2. IEEE, pp 994–1000

Baca T, Loianno G, Saska M (2016) Embedded model predictive control of unmanned micro aerial vehicles. In: Proceedings of 2016 21st international conference on methods and models in automation and robotics (MMAR). IEEE, pp 992–997

Bachmayer R, Humphris S, Fornari D, Van Dover C, Howland J, Bowen A, Elder R, Crook T, Gleason D, Sellers W, Lerner S (1998) Oceanographic research using remotely operated underwater robotic vehicles: exploration of hydrothermal vent sites on the mid-Atlantic ridge at 37 (degrees) north 32 (degrees) west. Marine Technol Soc J 32(3):37

Bellman R (1957) Dynamic programming. Princeton University Press, Princeton

Bingham B, Foley B, Singh H, Camilli R, Delaporta K, Eustice R, Mallios A, Mindell D, Roman C, Sakellariou D (2010) Robotic tools for deep water archaeology: surveying an ancient shipwreck with an autonomous underwater vehicle. J Field Rob 27(6):702–717

BlueRobotics (2022) The BlueROV2. https://bluerobotics.com/store/rov/bluerov2/. Accessed: 5 Aug 2022

Bryson AE, Ho YC (2018) Applied optimal control: optimization, estimation, and control. Routledge, London

Chen L, Hopman H, Negenborn RR (2018) Distributed model predictive control for vessel train formations of cooperative multi-vessel systems. Transp Res Part C Emerg Technol 92:101–118

Chen L, Hopman H, Negenborn RR (2019) Distributed model predictive control for cooperative floating object transport with multi-vessel systems. Ocean Eng 191:106515

Chisci L, Lombardi A, Mosca E (1996) Dual-receding horizon control of constrained discrete time systems. Euro J Control 4(2):278–285

de la Peña DM, Christofides PD (2008) Lyapunov-based model predictive control of nonlinear systems subject to data losses. IEEE Trans Autom Control 53(9):2076–2089

Desset S, Damus R, Hover F, Morash J, Polidoro V (2005) Closer to deep underwater science with ODYSSEY IV class hovering autonomous underwater vehicle (HAUV). In: Proceedings of Europe Oceans 2005, vol 2. IEEE, pp 758–762

do Nascimento AA, Feyzmahdavian HR, Johansson M, Garcia-Gabin W, Tervo K, (2019) Tube-based model predictive control for dynamic positioning of marine vessels. IFAC-Papers Online 52(21):33–38
Elvander J, Hawkes G (2012) ROVs and AUVs in support of marine renewable technologies. In: Proceedings of 2012 Oceans. IEEE, pp 1–6
Fernández DC, Hollinger GA (2016) Model predictive control for underwater robots in ocean waves. IEEE Robot Autom Lett 2(1):88–95
Ferranti L, Negenborn RR, Keviczky T, Alonso-Mora J (2018) Coordination of multiple vessels via distributed nonlinear model predictive control. In: Proceedings of the 2018 European control conference (ECC), Limassol, Cyprus. IEEE, pp 2523–2528
Fossen TI (2002) Marine control systems: guidance, navigation, and control of ships, rigs and underwater vehicles. In: Marine cybernetics, Trondheim, Norway
Fossen TI (2011) Handbook of marine craft hydrodynamics and motion control. Wiley, New Jersey
Gan W, Zhu D, Ji D (2018) Qpso-model predictive control-based approach to dynamic trajectory tracking control for unmanned underwater vehicles. Ocean Eng 158:208–220
Grüne L (2012) NMPC without terminal constraints. IFAC Proc 45(17):1–13
Guerreiro BJ, Silvestre C, Cunha R, Pascoal A (2014) Trajectory tracking nonlinear model predictive control for autonomous surface craft. IEEE Trans Control Syst Technol 22(6):2160–2175
Heshmati-Alamdari S, Karras GC, Marantos P, Kyriakopoulos KJ (2020) A robust predictive control approach for underwater robotic vehicles. IEEE Trans Control Syst Technol 28(6):2352–2363
Heshmati-Alamdari S, Nikou A, Dimarogonas DV (2020) Robust trajectory tracking control for underactuated autonomous underwater vehicles in uncertain environments. IEEE Trans Autom Sci Eng
Howard TM, Green CJ, Kelly A (2010) Receding horizon model-predictive control for mobile robot navigation of intricate paths. In: Field and service robotics. Springer, pp 69–78
Hung NT, Pascoal AM, Johansen TA (2020) Cooperative path following of constrained autonomous vehicles with model predictive control and event-triggered communications. Int J Robust Nonlinear Control 30(7):2644–2670
Huynh VT, Dunbabin M, Smith RN (2015) Predictive motion planning for AUVs subject to strong time-varying currents and forecasting uncertainties. In: Proceedings of the 2015 IEEE international conference on robotics and automation (ICRA). Seattle, USA. IEEE, pp 1144–1151
Jalbert J, Baker J, Duchesney J, Pietryka P, Dalton W, Blidberg DR, Chappell SG, Nitzel R, Holappa K (2003) Solar-powered autonomous underwater vehicle development. In: Proceedings of the 13th international symposium on unmanned untethered submersible technology, pp 1132–1140
Kalman RE (1960) Contributions to the theory of optimal control. Bol Soc Mat Mexicana 5(2):102–119
Kapetanović N, Bibuli M, Mišković N, Caccia M (2017) Real-time model predictive line following control for underactuated marine vehicles. IFAC-PapersOnLine 50(1):12374–12379
Keerthi SS, Gilbert EG (1988) Optimal infinite-horizon feedback laws for a general class of constrained discrete-time systems: stability and moving-horizon approximations. J Optim Theory Appl 57(2):265–293
Li H, Xie P, Yan W (2016) Receding horizon formation tracking control of constrained underactuated autonomous underwater vehicles. IEEE Trans Indus Electron 64(6):5004–5013
Li H, Yan W (2016) Model predictive stabilization of constrained underactuated autonomous underwater vehicles with guaranteed feasibility and stability. IEEE/ASME Trans Mechatron 22(3):1185–1194
Li Z, Sun J (2011) Disturbance compensating model predictive control with application to ship heading control. IEEE Trans Control Syst Technol 20(1):257–265
Li Z, Sun J, Oh S (2009) Path following for marine surface vessels with rudder and roll constraints: an MPC approach. In: Proceedings of the 2009 American control conference (ACC), St. Louis, MO, USA, . IEEE, pp 3611–3616
Liang H, Li H, Xu D (2020) and Demin Xu. Nonlinear model predictive trajectory tracking control of underactuated marine vehicles: theory and experiment. IEEE Trans Indus Electron 68(5):4238–48

Liu X, Shi Y, Constantinescu D (2014) Robust constrained model predictive control using contraction theory. In: Proceedings of 53rd IEEE conference on decision and control. IEEE, pp 3536–3541

Liu Z, Zhang Y, Yu X, Yuan C (2016) Unmanned surface vehicles: an overview of developments and challenges. Ann Rev Control 41:71–93

Saab Seaeye Ltd. (2022) The Falcon ROV. https://www.saabseaeye.com/solutions/underwater-vehicles/falcon

Mayne D (2013) An apologia for stabilising terminal conditions in model predictive control. Int J Control 86(11):2090–2095

Mayne DQ (2014) Model predictive control: recent developments and future promise. Automatica 50(12):2967–2986

Mayne DQ, Rawlings JB, Rao CV, Scokaert PO (2000) Constrained model predictive control: stability and optimality. Automatica 36(6):789–814

Meyer B, Ehlers K, Osterloh C, Maehle E (2013) Smart-E an autonomous omnidirectional underwater robot. Paladyn J Behav Robot 4(4):204–210

Michalska H (1997) A new formulation of receding horizon stabilising control without terminal constraint on the state. Euro J Control 3(1):2–14

Naeem W, Sutton R, Ahmad SM (2004) Pure pursuit guidance and model predictive control of an autonomous underwater vehicle for cable/pipeline tracking. In: Proceedings-institute of marine engineering science and technology part C journal of marine science and environment. Citeseer, pp 25–35

Oh SR, Sun J (2010) Path following of underactuated marine surface vessels using line-of-sight based model predictive control. Ocean Eng 37(2–3):289–295

Pavlov A, Nordahl H, Breivik M (2009) MPC-based optimal path following for underactuated vessels. IFAC Proc 42(18):340–345

Perez T, Tzeng CY, Goodwin GC (2000) Model predictive rudder roll stabilization control for ships. IFAC Proc 33(21):45–50

Primbs JA, Nevistic V (1997) Constrained finite receding horizon linear quadratic control. 4:3196–3201

Qin SJ, Badgwell TA (2000) An overview of nonlinear model predictive control applications. In: Nonlinear model predictive control, pp 369–392

Rawlings JB, Mayne DQ, Diehl M (2017) Model predictive control: theory, computation, and design, vol 2. Nob Hill Publishing, Madison

Ribas D, Ridao P, Neira J (2010) Underwater SLAM for structured environments using an imaging sonar, vol 65. Springer, New York

Roberts GN, Sutton R (2006) Advances in unmanned marine vehicles, vol 69. The Institution of Electrical Engineers, Stevenage

Maritime Robotics (2022) The Otter autonomous surface vehicle. https://www.maritimerobotics.com/otter. Accessed: 08 May 2022

Roman C, Mather R (2010) Autonomous underwater vehicles as tools for deep-submergence archaeology. In: Proceedings of the institution of mechanical engineers. Part M J Eng Marit Environ 224(4):327–340

Scattolini R (2009) Architectures for distributed and hierarchical model predictive control-a review. J Process Control 19(5):723–731

Scokaert PO, Mayne DQ, Rawlings JB (1999) Suboptimal model predictive control (feasibility implies stability). IEEE Trans Autom Control 44(3):648–654

Shen C, Buckham B, Shi Y (2016) Modified C/GMRES algorithm for fast nonlinear model predictive tracking control of AUVs. IEEE Trans Control Syst Technol 25(5):1896–1904

Shen C, Shi Y (2020) Distributed implementation of nonlinear model predictive control for AUV trajectory tracking. Automatica 115:108863

Shen C, Shi Y, Buckham B (2016) Integrated path planning and tracking control of an AUV: a unified receding horizon optimization approach. IEEE/ASME Trans Mechatron 22(3):1163–1173

Shen C, Shi Y, Buckham B (2017) Integrated path planning and tracking control of an AUV: a unified receding horizon optimization approach. IEEE/ASME Trans Mechatron 22(3):1163–1173
Shen C, Shi Y, Buckham B (2017) Lyapunov-based model predictive control for dynamic positioning of autonomous underwater vehicles. In: Proceedings of 2017 IEEE international conference on unmanned systems (ICUS). IEEE, pp 588–593
Shen C, Shi Y, Buckham B (2017) Trajectory tracking control of an autonomous underwater vehicle using Lyapunov-based model predictive control. IEEE Trans Indus Electron 65(7):5796–5805
Shen C, Shi Y, Buckham B (2018) Path-following control of an AUV: a multiobjective model predictive control approach. IEEE Trans Control Syst Technol 27(3):1334–1342
Shi Y, Shen C, Fang H, Li H (2017) Advanced control in marine mechatronic systems: a survey. IEEE/ASME Trans Mechatron 22(3):1121–1131
Sørensen AJ (2011) A survey of dynamic positioning control systems. Ann Rev Control 35(1):123–136
Tsopelakos A, Papadopoulos E (2016) Design and evaluation of dynamic positioning controllers with parasitic thrust reduction for an overactuated floating platform. IEEE Trans Control Syst Technol 25(1):145–160
Veksler A, Johansen TA, Borrelli F, Realfsen B (2016) Dynamic positioning with model predictive control. IEEE Trans Control Syst Technol 24(4):1340–1353
Wahl A, Gilles ED (1998) Track-keeping on waterways using model predictive control. IFAC Proc 31(30):149–154
Wei H, Shen C, Shi Y (2021) Distributed Lyapunov-based model predictive formation tracking control for autonomous underwater vehicles subject to disturbances. IEEE Trans Syst Man Cybern Syst 51(8):5198–5208
Wei H, Sun Q, Chen J, Shi Y (2021) Robust distributed model predictive platooning control for heterogeneous autonomous surface vehicles. Control Eng Pract 107:104655
Wernli RL (2000) AUV commercialization-who's leading the pack? In: Proceedings of Oceans 2000 MTS/IEEE conference and exhibition, vol 1. IEEE, pp 391–395
Wynn RB, Huvenne VA, Le Bas TP, Murton BJ, Connelly DP, Bett BJ, Ruhl HA, Morris KJ, Peakall J, Parsons DR (2014) Autonomous underwater vehicles (AUVs): their past, present and future contributions to the advancement of marine geoscience. Marine Geol 352:451–468
Yan Z, Wang J (2012) Model predictive control for tracking of underactuated vessels based on recurrent neural networks. IEEE J Oceanic Eng 37(4):717–726
Yang Y, Wang Y, Manzie C, Pu Y (2021) Real-time distributed MPC for multiple underwater vehicles with limited communication data-rates. IEEE Proceedings of 2021 American Control Conference (ACC), pp 3314–3319
Yoerger DR, Jakuba M, Bradley AM, Bingham B (2007) Techniques for deep sea near bottom survey using an autonomous underwater vehicle. Int J Robot Res 26(1):41–54
Yoon S, Qiao C (2010) Cooperative search and survey using autonomous underwater vehicles (AUVs). IEEE Trans Parallel Distrib Syst 22(3):364–379
Yuh J (2000) Design and control of autonomous underwater robots: a survey. Auton Robot 8(1):7–24
Zhang J, Sun T, Liu Z (2017) Robust model predictive control for path-following of underactuated surface vessels with roll constraints. Ocean Eng 143:125–132
Zhang Y, Liu X, Luo M, Yang C (2019) MPC-based 3-D trajectory tracking for an autonomous underwater vehicle with constraints in complex ocean environments. Ocean Eng 189:106309
Zheng H, Negenborn RR, Lodewijks G (2016) Fast ADMM for distributed model predictive control of cooperative waterborne AGVs. IEEE Trans Control Syst Technol 25(4):1406–1413
Zheng H, Negenborn RR, Lodewijks G (2017) Robust distributed predictive control of waterborne AGVs-a cooperative and cost-effective approach. IEEE Trans Cybernet 48(8):2449–2461
Zheng H, Wu J, Wu W, Zhang Y (2020) Robust dynamic positioning of autonomous surface vessels with tube-based model predictive control. Ocean Eng 199:106820

Chapter 2
AMV Modeling

2.1 Kinematics of AMVs

2.1.1 Reference Frames

The sensors provide their measurements with respect to different reference frames. When studying the motion control problems, it is convenient to use two reference frames (see Fig. 2.1) to describe the AMV motion state and the control objective:

- The body-fixed reference frame (BRF) is affixed to the vehicle with the origin selected to be the center of gravity (CG). The body axes are defined such that they coincide with the principal axes of inertia: The longitude axis which points from aft to fore is often referred to as the x_b axis; the transversal axis which points from port to starboard is the y_b axis; the z_b axis is defined orthogonal to both x_b and y_b axes and obeys the right-hand rule.
- Then the motion of the AMV can be described as the BRF motion relative to an inertial reference frame (IRF) which is used to record the footprints of the vehicle and to specify the control objectives. Usual selections of IRF include the Earth-centered inertial (ECI) frame, the Earth-centered Earth-fixed (ECEF) reference frame, and the North–East–Down (NED) coordinate system (Fossen 2002).

The linear and angular velocities of the vehicle are expressed in the BRF while the position and orientation are described with respect to the IRF. The vectorial forms of these expressions are as follows:

- $\boldsymbol{\eta} = [x, y, z, \phi, \theta, \psi]^{\mathrm{T}}$, the position and orientation vector represented in IRF
- $\mathbf{v} = [u, v, w, p, q, r]^{\mathrm{T}}$, the velocity vector represented in BRF.

Since in many navigation applications, the position vector is decomposed in NED coordinates, the IRF is selected to be coincident with the NED coordinate system in this book.

Y. Shi et al., *Advanced Model Predictive Control for Autonomous Marine Vehicles*, Advances in Industrial Control, https://doi.org/10.1007/978-3-031-19354-5_2

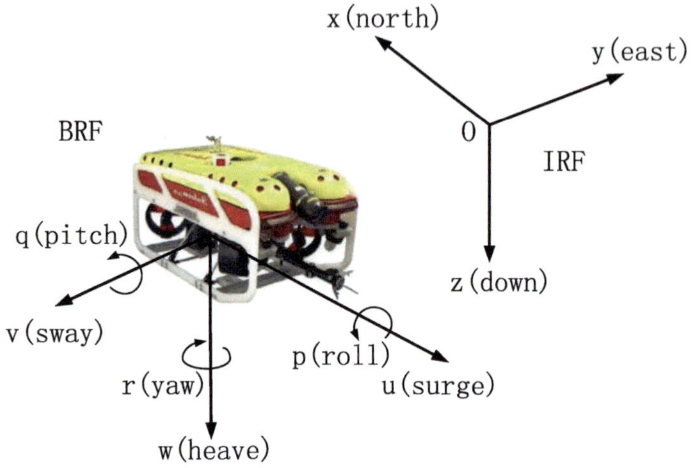

Fig. 2.1 Illustration of the reference frames for the AUV motion control. Photo courtesy of Saab Seaeye Ltd. © 2022 Saab Seaeye Ltd., reproduced with permission from (Saab Seaeye Ltd. 2022)

2.1.2 Transformation Between Reference Frames

Rotation matrices are essential in deriving the kinematic equations of motion for an AMV. The rotation matrix between the BRF and the IRF is denoted as $\mathbf{R}_b^i$ which belongs to the special orthogonal group of order three SO(3):

$$\mathrm{SO}(3) = \{\mathbf{R} \mid \mathbf{R} \in \mathbb{R}^{3\times 3},\ \mathbf{R}\mathbf{R}^{\mathrm{T}} = \mathbf{R}^{\mathrm{T}}\mathbf{R} = \mathbf{I},\ \det \mathbf{R} = 1\}. \tag{2.1}$$

Let $\mathbf{v}_o^b = [u, v, w]^{\mathrm{T}}$ denote the linear velocity vector fixed in BRF and $\mathbf{v}_o^i$ denote this velocity vector decomposed in IRF. Then the relation between them can be expressed using the following equation:

$$\mathbf{v}_o^i = \mathbf{R}_b^i(\Theta)\mathbf{v}_o^b, \tag{2.2}$$

where $\Theta = [\phi, \theta, \psi]^{\mathrm{T}}$ encloses the roll angle (ϕ), pitch angle (θ), and yaw angle (ψ). In navigation, guidance, and control applications, the zyx-convention is commonly adopted to describe the rotation matrix $\mathbf{R}_b^i(\Theta)$:

$$\mathbf{R}_b^i(\Theta) = \mathbf{R}_{z,\psi}\mathbf{R}_{y,\theta}\mathbf{R}_{x,\phi}, \tag{2.3}$$

with

$$\mathbf{R}_{x,\phi} = \begin{bmatrix} 1 & 0 & 0 \\ 0 & c\phi & -s\phi \\ 0 & s\phi & c\phi \end{bmatrix},\ \mathbf{R}_{y,\theta} = \begin{bmatrix} c\theta & 0 & s\theta \\ 0 & 1 & 0 \\ -s\theta & 0 & c\theta \end{bmatrix},\ \mathbf{R}_{z,\psi} = \begin{bmatrix} c\psi & -s\psi & 0 \\ s\psi & c\psi & 0 \\ 0 & 0 & 1 \end{bmatrix}.$$

Here, $s\cdot$, $c\cdot$ are shorthand for trigonometric functions $\sin(\cdot)$, $\cos(\cdot)$. The inverse transformation satisfies

$$\mathbf{R}_b^i(\Theta)^{-1} = \mathbf{R}_i^b(\Theta) = \mathbf{R}_{x,\phi}^{\mathrm{T}}\mathbf{R}_{y,\theta}^{\mathrm{T}}\mathbf{R}_{z,\psi}^{\mathrm{T}}. \tag{2.4}$$

Expanding (2.3), we have

$$\mathbf{R}_b^i(\Theta) = \begin{bmatrix} c\psi c\theta & c\psi s\theta s\phi - s\psi c\phi & c\psi c\theta s\phi + s\psi s\phi \\ s\psi c\theta & s\psi s\theta c\phi + c\psi c\phi & s\psi s\theta c\phi - c\psi s\phi \\ -s\theta & c\theta s\phi & c\theta c\phi \end{bmatrix}. \tag{2.5}$$

Let $\boldsymbol{\omega}_{ib}^b = [p, q, r]^{\mathrm{T}}$ denote the angular velocity of BRF relative to IRF decomposed in BRF and $\dot{\Theta} = [\dot{\phi}, \dot{\theta}, \dot{\psi}]$ denote the Euler-angle rate. They are related by the following equation:

$$\dot{\Theta} = \mathbf{T}(\Theta)\boldsymbol{\omega}_{ib}^b. \tag{2.6}$$

The transformation matrix $\mathbf{T}(\Theta)$ can be derived through the following relation:

$$\boldsymbol{\omega}_{ib}^b = \begin{bmatrix} \dot{\phi} \\ 0 \\ 0 \end{bmatrix} + \mathbf{R}_{x,\phi}^{\mathrm{T}} \begin{bmatrix} 0 \\ \dot{\theta} \\ 0 \end{bmatrix} + \mathbf{R}_{x,\phi}^{\mathrm{T}}\mathbf{R}_{y,\theta}^{\mathrm{T}} \begin{bmatrix} 0 \\ 0 \\ \dot{\psi} \end{bmatrix}. \tag{2.7}$$

Expanding (2.7) we have

$$\mathbf{T}(\Theta) = \begin{bmatrix} 1 & s\phi t\theta & c\phi t\theta \\ 0 & c\phi & -s\phi \\ 0 & s\phi/c\theta & c\phi/c\theta \end{bmatrix}, \tag{2.8}$$

where $t\cdot$ is shorthand for $\tan(\cdot)$. Combining (2.2) and (2.6), we can have the 6-DOF kinematic equations of motion expressed in the vectorial form

$$\dot{\boldsymbol{\eta}} = \begin{bmatrix} \dot{\mathbf{p}} \\ \dot{\Theta} \end{bmatrix} = \begin{bmatrix} \mathbf{R}_b^i(\Theta) & \mathbf{0}_{3\times3} \\ \mathbf{0}_{3\times3} & \mathbf{T}(\Theta) \end{bmatrix} \begin{bmatrix} \mathbf{v}_o^b \\ \boldsymbol{\omega}_{ib}^b \end{bmatrix} = \mathbf{J}(\boldsymbol{\eta})\mathbf{v}, \tag{2.9}$$

where $\mathbf{p} = [x, y, z]^{\mathrm{T}}$ is the position vector of the vehicle represented in IRF.

2.2 Nonlinear Dynamics of AMVs

2.2.1 Rigid-Body Dynamics

To facilitate the derivation of the dynamic equations of AMV motion, it is common and reasonable to assume that the vehicle is a rigid body, which eliminates the need of analyzing the interactions between individual elements of mass.

The rigid-body dynamics of the AMV can be derived by applying the Newtonian mechanics (Fossen 2002):

$$\mathbf{M}_{\mathrm{RB}}\dot{\mathbf{v}} + \mathbf{C}_{\mathrm{RB}}(\mathbf{v})\mathbf{v} = \boldsymbol{\tau}_{\mathrm{RB}}, \tag{2.10}$$

where $\boldsymbol{\tau}_{\mathrm{RB}} = [X, Y, Z, K, M, N]^{\mathrm{T}}$ is the generalized external force and moment vector expressed in BRF. Since the origin of BRF is coincident with the CG of AMV, the rigid-body inertia matrix $\mathbf{M}_{\mathrm{RB}}$ can be simplified as

$$\mathbf{M}_{\mathrm{RB}} = \begin{bmatrix} m\mathbf{I}_{3\times 3} & \mathbf{0} \\ \mathbf{0} & \mathbf{I}_o \end{bmatrix}, \tag{2.11}$$

where m is the mass of the vehicle and $\mathbf{I}_o$ is the inertia tensor defined as

$$\mathbf{I}_o = \begin{bmatrix} I_x & -I_{xy} & -I_{xz} \\ -I_{yx} & I_y & -I_{yz} \\ -I_{zx} & -I_{zy} & I_z \end{bmatrix}, \tag{2.12}$$

and the rigid-body Coriolis and centripetal matrix $\mathbf{C}_{\mathrm{RB}}$ is

$$\mathbf{C}_{\mathrm{RB}}(\mathbf{v}) = \begin{bmatrix} \mathbf{0}_{3\times 3} & -m\mathbf{S}(\mathbf{v}_o^b) \\ -m\mathbf{S}(\mathbf{v}_o^b) & -\mathbf{S}(\mathbf{I}_0\boldsymbol{\omega}_{ib}^b) \end{bmatrix}, \tag{2.13}$$

where $\mathbf{S}(\cdot)$ is the cross-product operator.

Definition 2.1 (Cross-Product Operator) The cross-product of two vectors $\mathbf{a} \times \mathbf{b}$ can be expressed using normal matrix multiplication:

$$\mathbf{a} \times \mathbf{b} = \mathbf{S}(\mathbf{a})\mathbf{b}, \tag{2.14}$$

and the operator $\mathbf{S}(\cdot)$ is defined as

$$\mathbf{S}(\mathbf{a}) = -\mathbf{S}^{\mathrm{T}}(\mathbf{a}) = \begin{bmatrix} 0 & -a_3 & a_2 \\ a_3 & 0 & -a_1 \\ -a_2 & a_1 & 0 \end{bmatrix}, \quad \mathbf{a} = \begin{bmatrix} a_1 \\ a_2 \\ a_3 \end{bmatrix}. \tag{2.15}$$

2.2.2 *Hydrodynamic Forces and Moments*

The hydrodynamics should be considered in calculating the total external forces and moments $\boldsymbol{\tau}_{\mathrm{RB}}$. Several main contributions of the hydrodynamic forces and moments include the radiation-induced forces, skin friction damping, wave drift damping, damping due to vortex shedding and environmental disturbances. They are treated separately based on the principle of superposition.

The radiation-induced forces and moments include three components, namely added mass, potential damping, and restoring forces. They can be expressed mathematically as follows:

$$\boldsymbol{\tau}_R = -\mathbf{M}_A\dot{\mathbf{v}} - \mathbf{C}_A(\mathbf{v})\mathbf{v} - \mathbf{D}_P(\mathbf{v})\mathbf{v} - \mathbf{g}(\boldsymbol{\eta}), \tag{2.16}$$

where $-\mathbf{M}_A\dot{\mathbf{v}} - \mathbf{C}_A(\mathbf{v})\mathbf{v}$ is the added mass term, $-\mathbf{D}_P(\mathbf{v})\mathbf{v}$ is the potential damping term, and $-\mathbf{g}(\boldsymbol{\eta})$ is the restoring force term. The inertia matrix of the added mass is defined as

$$\mathbf{M}_A = \begin{bmatrix} X_{\dot{u}} & X_{\dot{v}} & X_{\dot{w}} & X_{\dot{p}} & X_{\dot{q}} & X_{\dot{r}} \\ Y_{\dot{u}} & Y_{\dot{v}} & Y_{\dot{w}} & Y_{\dot{p}} & Y_{\dot{q}} & Y_{\dot{r}} \\ Z_{\dot{u}} & Z_{\dot{v}} & Z_{\dot{w}} & Z_{\dot{p}} & Z_{\dot{q}} & Z_{\dot{r}} \\ K_{\dot{u}} & K_{\dot{v}} & K_{\dot{w}} & K_{\dot{p}} & K_{\dot{q}} & K_{\dot{r}} \\ M_{\dot{u}} & M_{\dot{v}} & M_{\dot{w}} & M_{\dot{p}} & M_{\dot{q}} & M_{\dot{r}} \\ N_{\dot{u}} & N_{\dot{v}} & N_{\dot{w}} & N_{\dot{p}} & N_{\dot{q}} & N_{\dot{r}} \end{bmatrix}, \tag{2.17}$$

where the hydrodynamic coefficients are defined as partial derivative of the added mass force over the corresponding acceleration. For example, the added mass force X_A along the x-axis due to the acceleration w is $X_{A1} = X_{\dot{w}}\dot{w}$, and $X_{\dot{w}} = \partial X_A/\partial \dot{w}$.

The hydrodynamic Coriolis and centripetal matrix C_A can be calculated using the following formula (Fossen 2002):

$$\mathbf{C}_A(\mathbf{v}) = \begin{bmatrix} \mathbf{0}_{3\times3} & -\mathbf{S}(\mathbf{A}_{11}\mathbf{v}_o^b + \mathbf{A}_{12}\boldsymbol{\omega}_{ib}^b) \\ -\mathbf{S}(\mathbf{A}_{11}\mathbf{v}_o^b + \mathbf{A}_{12}\boldsymbol{\omega}_{ib}^b) & -\mathbf{S}(\mathbf{A}_{21}\mathbf{v}_o^b + \mathbf{A}_{22}\boldsymbol{\omega}_{ib}^b) \end{bmatrix}, \tag{2.18}$$

where $\mathbf{A} = \mathbf{A}^{\mathrm{T}}$ defined as

$$\mathbf{A} = \frac{1}{2}(\mathbf{M}_A + \mathbf{M}_A^{\mathrm{T}}),\ \mathbf{A} = \begin{bmatrix} \mathbf{A}_{11} & \mathbf{A}_{12} \\ \mathbf{A}_{21} & \mathbf{A}_{22} \end{bmatrix},\ \mathbf{A}_{ij} \in \mathbb{R}^{3\times3}. \tag{2.19}$$

In addition to potential damping the skin friction damping, wave drift damping, damping due to vortex shedding need to be included, and those damping forces and moments can be expressed as

$$\boldsymbol{\tau}_D = -\mathbf{D}_S(\mathbf{v})\mathbf{v} - \mathbf{D}_W(\mathbf{v})\mathbf{v} - \mathbf{D}_M(\mathbf{v})\mathbf{v}. \tag{2.20}$$

Define the total hydrodynamic damping matrix as

$$\mathbf{D}(\mathbf{v}) = \mathbf{D}_P(\mathbf{v}) + \mathbf{D}_S(\mathbf{v}) + \mathbf{D}_W(\mathbf{v}) + \mathbf{D}_M(\mathbf{v}), \tag{2.21}$$

then we have the hydrodynamic force and moment vector $\boldsymbol{\tau}_H$ written as the sum of $\boldsymbol{\tau}_R$ and $\boldsymbol{\tau}_D$, i.e.,

$$\boldsymbol{\tau}_H = -\mathbf{M}_A\dot{\mathbf{v}} - \mathbf{C}_A(\mathbf{v})\mathbf{v} - \mathbf{D}(\mathbf{v})\mathbf{v} - \mathbf{g}(\boldsymbol{\eta}). \tag{2.22}$$

The total hydrodynamic damping forces are composed of linear damping terms and quadratic damping terms and can be conveniently expressed as

$$\mathbf{D}(\mathbf{v})\mathbf{v} = \mathbf{D}_L\mathbf{v} + \begin{bmatrix} |\mathbf{v}|^{\mathrm{T}}\mathbf{D}_{n1}\mathbf{v} \\ |\mathbf{v}|^{\mathrm{T}}\mathbf{D}_{n2}\mathbf{v} \\ |\mathbf{v}|^{\mathrm{T}}\mathbf{D}_{n3}\mathbf{v} \\ |\mathbf{v}|^{\mathrm{T}}\mathbf{D}_{n4}\mathbf{v} \\ |\mathbf{v}|^{\mathrm{T}}\mathbf{D}_{n5}\mathbf{v} \\ |\mathbf{v}|^{\mathrm{T}}\mathbf{D}_{n6}\mathbf{v} \end{bmatrix}, \tag{2.23}$$

where $\mathbf{D}_L$ is the linear damping matrix, and $\mathbf{D}_{ni}$, $i = 1, 2, \ldots, 6$ are quadratic damping matrices. The restoring forces and moments are calculated as follows:

$$\mathbf{g}(\boldsymbol{\eta}) = \begin{bmatrix} (W - B)s\theta \\ -(W - B)c\theta s\phi \\ -(W - B)c\theta c\phi \\ y_B Bc\theta c\phi - z_B Bc\theta s\phi \\ -z_B Bs\theta - x_B Bc\theta c\phi \\ x_B Bc\theta s\phi + y_B Bs\theta \end{bmatrix}, \tag{2.24}$$

where $W = mg$ is the gravity, $B = bg$ is the buoyancy, and $[x_B, y_B, z_B]$ is the coordinates of the center of buoyancy with respect to BRF.

Let $\mathbf{w}$ denote the environmental disturbances which exist due to waves and ocean currents. The total external forces and moments $\boldsymbol{\tau}_{\mathrm{RB}}$ can be expressed as

$$\boldsymbol{\tau}_{\mathrm{RB}} = \boldsymbol{\tau}_H + \mathbf{w} + \boldsymbol{\tau}, \tag{2.25}$$

where $\boldsymbol{\tau}$ represents the propulsive forces and moments. Then we can have the 6-DOF dynamic equations of motion arranged in the following form:

$$\mathbf{M}\dot{\mathbf{v}} + \mathbf{C}(\mathbf{v})\mathbf{v} + \mathbf{D}(\mathbf{v})\mathbf{v} + \mathbf{g}(\boldsymbol{\eta}) = \boldsymbol{\tau} + \mathbf{w}, \tag{2.26}$$

where

$$\mathbf{M} = \mathbf{M}_{\mathrm{RB}} + \mathbf{M}_A, \quad \mathbf{C}(\mathbf{v}) = \mathbf{C}_{\mathrm{RB}}(\mathbf{v}) + \mathbf{C}_A(\mathbf{v}). \tag{2.27}$$

2.3 AMV Model for Motion Control Design

In this book, we consider the motion of the AMV in the local level plane. Three mild assumptions can be satisfied for the low-speed motion of AMV: (i) The vehicle is with three planes of symmetry; (ii) the mass distribution is homogeneous; and (iii) the pitch and roll motions are neglected. As a result, for the motion control in the local level plane, the system matrices in (2.26) can be simplified. The inertia matrix becomes

$$\mathbf{M} = \begin{bmatrix} M_{\dot{u}} & 0 & 0 \\ 0 & M_{\dot{v}} & 0 \\ 0 & 0 & M_{\dot{r}} \end{bmatrix}, \tag{2.28}$$

where $M_{\dot{u}} = m - X_{\dot{u}}$, $M_{\dot{v}} = m - Y_{\dot{v}}$, and $M_{\dot{r}} = I_z - N_{\dot{r}}$ are the inertia terms including add mass. The restoring force is neglected $\mathbf{g}(\boldsymbol{\eta}) = \mathbf{0}$, and the damping matrix is

$$\mathbf{D}(\mathbf{v}) = \begin{bmatrix} X_u + D_u|u| & 0 & 0 \\ 0 & Y_v + D_v|v| & 0 \\ 0 & 0 & N_r + D_r|r| \end{bmatrix}, \tag{2.29}$$

where X_u, Y_v, N_r are linear drag coefficients, and D_u, D_v, D_r are the quadratic drag coefficients. The Coriolis and centripetal matrix becomes

$$\mathbf{C}(\mathbf{v}) = \begin{bmatrix} 0 & 0 & -M_{\dot{v}}v \\ 0 & 0 & M_{\dot{u}}u \\ M_{\dot{v}}v & -M_{\dot{u}}u & 0 \end{bmatrix}. \tag{2.30}$$

In the local level plane, the velocity vector $\mathbf{v} = [u, v, r]^T$ encloses the surge, sway, and yaw velocities, and the position and orientation vector $\boldsymbol{\eta} = [x, y, \psi]^T$ includes the position and heading of the vehicle.

In the AMV motion controller design, we assume that the disturbances are small, i.e., $\mathbf{w} \approx \mathbf{0}$. Then the dynamic equations of motion under consideration is:

$$\mathbf{M}\dot{\mathbf{v}} + \mathbf{C}(\mathbf{v})\mathbf{v} + \mathbf{D}(\mathbf{v})\mathbf{v} + \mathbf{g}(\boldsymbol{\eta}) = \boldsymbol{\tau}, \tag{2.31}$$

where $\boldsymbol{\tau} = [F_u, F_v, F_r]^T$ denotes the generalized thrust forces and moments. Further expanding the dynamic equations (2.31) into the elementwise expression, we have the following equations:

$$\dot{u} = \frac{M_{\dot{v}}}{M_{\dot{u}}}vr - \frac{X_u}{M_{\dot{u}}}u - \frac{D_u}{M_{\dot{u}}}u|u| + \frac{F_u}{M_{\dot{u}}}, \tag{2.32a}$$

$$\dot{v} = -\frac{M_{\dot{u}}}{M_{\dot{v}}}ur - \frac{Y_v}{M_{\dot{v}}}v - \frac{D_v}{M_{\dot{v}}}v|v| + \frac{F_v}{M_{\dot{v}}}, \tag{2.32b}$$

$$\dot{r} = \frac{M_{\dot{u}} - M_{\dot{v}}}{M_{\dot{r}}}uv - \frac{N_r}{M_{\dot{r}}}r - \frac{D_r}{M_{\dot{r}}}r|r| + \frac{F_r}{M_{\dot{r}}}. \tag{2.32c}$$

The kinematic equations (2.9) can also be simplified as follows:

$$\dot{\boldsymbol{\eta}} = \begin{bmatrix} \cos\psi & -\sin\psi & 0 \\ \sin\psi & \cos\psi & 0 \\ 0 & 0 & 1 \end{bmatrix} \begin{bmatrix} u \\ v \\ r \end{bmatrix} = \mathbf{R}(\psi)\mathbf{v}. \tag{2.33}$$

Table 2.1 Hydrodynamic coefficient summary

Inertia term	Linear drag	Quadratic drag
$M_{\dot{u}} = 283.6$ kg	$X_u = 26.9$ kg/s	$D_u = 241.3$ kg/m
$M_{\dot{v}} = 593.2$ kg	$Y_v = 35.8$ kg/s	$D_v = 503.8$ kg/m
$M_{\dot{r}} = 29.0$ kg m^2	$N_r = 3.5$ kg m^2/s	$D_r = 76.9$ kg m^2

Further expanding the kinematic equations (2.33) into the elementwise expression, we have

$$\dot{x} = u\cos\psi - v\sin\psi, \tag{2.34a}$$

$$\dot{y} = u\sin\psi + v\cos\psi, \tag{2.34b}$$

$$\dot{\psi} = r. \tag{2.34c}$$

Define the system state as $\mathbf{x} = [\boldsymbol{\eta}^{\mathrm{T}}, \mathbf{v}^{\mathrm{T}}]^{\mathrm{T}}$ and the generalized control input as $\boldsymbol{\tau}$. From (2.9) and (2.31), we obtain the general form of the AMV model

$$\dot{\mathbf{x}} = \begin{bmatrix} \mathbf{R}(\psi)\mathbf{v} \\ \mathbf{M}^{-1}(\boldsymbol{\tau} - \mathbf{C}(\mathbf{v})\mathbf{v} - \mathbf{D}(\mathbf{v})\mathbf{v} - \mathbf{g}(\boldsymbol{\eta})) \end{bmatrix} = \bar{\mathbf{f}}(\mathbf{x}, \boldsymbol{\tau}). \tag{2.35}$$

The generalized control input $\boldsymbol{\tau}$ is the resulting force of the thrusters. For the Falcon AUV, the experimental platform, four thrusters are effective in the local level plane. The relationship between them is described by the following thrust distribution function:

$$\boldsymbol{\tau} = \mathbf{B}\mathbf{u}, \tag{2.36}$$

where $\mathbf{u} = [u_1, u_2, u_3, u_4]^{\mathrm{T}}$ denotes the force provided by each thruster; $\mathbf{B}$ is the input matrix. Then the control system model used for the AMV motion control can be established by simply binding the kinematic equations, dynamics equations, and the thrust distribution function

$$\dot{\mathbf{x}} = \begin{bmatrix} \mathbf{R}(\psi)\mathbf{v} \\ \mathbf{M}^{-1}(\mathbf{B}\mathbf{u} - \mathbf{C}(\mathbf{v})\mathbf{v} - \mathbf{D}(\mathbf{v})\mathbf{v} - \mathbf{g}(\boldsymbol{\eta})) \end{bmatrix} = \mathbf{f}(\mathbf{x}, \mathbf{u}). \tag{2.37}$$

The hydrodynamic coefficients for the Falcon AUV (2.37) are summarized in Table 2.1 which are extracted from the previous modeling experiments based on (Proctor 2014). Then the input matrix is

$$\mathbf{B} = \begin{bmatrix} 0.7974 & 0.8643 & 0.8127 & 0.8270 \\ 0.6032 & 0.5029 & -0.5824 & -0.5610 \\ 0.2945 & -0.3302 & -0.2847 & 0.3505 \end{bmatrix}. \tag{2.38}$$

For the established AMV model (2.37), the following important properties can be easily explored and will be exploited in the controller design:

- P-1: The inertia matrix is positive definite and upper bounded: $0 < \mathbf{M} = \mathbf{M}^{\mathrm{T}} \leq \bar{m}\mathbf{I} < \infty$.
- P-2: The Coriolis and centripetal matrix is skew-symmetric: $\mathbf{C}(\mathbf{v}) = -\mathbf{C}^{\mathrm{T}}(\mathbf{v})$.
- P-3: The inverse of rotation matrix satisfies: $\mathbf{R}^{-1}(\psi) = \mathbf{R}^{\mathrm{T}}(\psi)$, and it preserves length $\|\mathbf{R}^{\mathrm{T}}(\psi)\dot{\boldsymbol{\eta}}\|_2 = \|\dot{\boldsymbol{\eta}}\|_2$.
- P-4: The damping matrix is positive definite: $\mathbf{D}(\mathbf{v}) > 0$.
- P-5: The input matrix satisfies that $\mathbf{B}\mathbf{B}^{\mathrm{T}}$ is nonsingular.
- P-6: The restoring force $\mathbf{g}(\boldsymbol{\eta})$ is bounded: $\|\mathbf{g}(\boldsymbol{\eta})\|_{\infty} \leq \bar{g}$.

2.4 Notes and Summary

In this chapter, we discussed the kinematics and dynamics of the AMV motion. The AMV mathematical model for the experimental platform was established based on the kinematic equations of motion, the dynamic equations of motion, and the thrust distribution function. The hydrodynamic coefficients were provided, and several important model properties that would be exploited in the motion controller design were explored.

References

Fossen TI (2002) Marine control systems: guidance, navigation, and control of ships, rigs and underwater vehicles. Marine Cybern Trondheim Norway

Proctor AA (2014) Semi-autonomous guidance and control of a Saab SeaEye Falcon ROV. Ph.D thesis

Saab Seaeye Ltd. (2022) The Falcon ROV. https://www.saabseaeye.com/solutions/underwater-vehicles/falcon

Chapter 3
Receding Horizon Optimization for Integrated Path Planning and Tracking Control of an AUV

3.1 Introduction

3.1.1 Research Background and Contributions

Trajectory tracking, being a basic robotic control problem, has been extensively studied for AUVs in the past several decades. For the tracking of piecewise linear paths, the line-of-sight (LOS) scheme is often used (Fossen and Pettersen 2014). To stabilize the cross-track error in the LOS scheme, conventional proportional–integral–derivative (PID) control (Fossen 2002), linear quadratic Gaussian (LQG) (Holzhüter 1997), and nonlinear PID techniques (Lindegaard 2022) have been applied to AUVs. For tracking of time-parameterized curves, the Lyapunov-based backstepping technique can be applied (Repoulias and Papadopoulos 2007). Due to the insensitivity to parametric uncertainty, the sliding mode control (Wang et al. 2012) is suitable for the AUV tracking control as well. However, the aforementioned control methods cannot handle the practical system constraints such as actuator limits. This motivates control theorists and AUV practitioners to investigate the MPC for the AUV trajectory-tracking problem. The beauty of MPC lies in the fact that it can conveniently handle nonlinear MIMO system control problems and explicitly take system constraints into consideration (Mayne et al. 2000). Linear MPC formulation of AUV tracking control has been investigated based on the linearization of the AUV model (Naeem et al. 2004). Linear MPC inherits the merit of the convex optimization problem which can be efficiently solved by off-the-shelf NLP algorithms (Antoniou and Lu 2007) to achieve real-time control. On the other hand, the demerit of this configuration is drawn from the approximation error of linearization. Naturally, the results in Linear MPC tracking control can be extended to NMPC formulation with the sacrifice of algorithm efficiency for model accuracy (Caldwell et al. 2010). Nevertheless, among all of the existing results of MPC tracking control, they either only take the kinematic equations as the vehicle model, (Gu and Hu 2006; Caldwell et al. 2010), or consider both kinematics and dynamics, but no rigorous proof of stability

Y. Shi et al., *Advanced Model Predictive Control for Autonomous Marine Vehicles*, Advances in Industrial Control, https://doi.org/10.1007/978-3-031-19354-5_3

is conducted in the controller design (Naeem et al. 2004). This motivates the main objective of the work in this chapter: To design an NMPC-based tracking controller considering both dynamics and kinematics of the AUV and provide the stability analysis for the closed-loop control system.

Path planning plays an essential role in tracking control. The conventional marine vessel guidance system (Fossen 2002) which generates whole-journey way-points does not work for AUVs in many cases since the global oceanic information may not be available *a priori*. Instead, the fast, reactive, and dynamic path planning methods are desirable in real-world scenarios (Yu et al. 2014; Jung et al. 2015). Recently, a spline-based path planning method is reported in (Berglund et al. 2010) for autonomous mining vehicle applications. In view of the similar essentials of autonomous vehicles, the basic principle is probably applicable to AUVs as well. The global environment information is assumed to be known in (Berglund et al. 2010). Considering the practical operation of AUVs with a limited sensing range, the second objective of the work in this chapter appears: To accommodate the spline-based path planning method to the limited perceiving capability of AUVs in the tracking control application.

In this chapter, by introducing a virtual reference system, the trajectory-tracking control problem can be transformed into the regulation problem of the error dynamics. The optimal value function of the associated OCP solved at each sampling instant can be shown nonincreasing along the trajectory of the closed-loop system, which guarantees the asymptotic stability of the closed-loop system. Furthermore, a dynamic path planning method is developed based on the RHO, enabling the natural integration of path planning and tracking control.

The main contributions of this chapter are threefold:

- A novel RHO formulation is proposed for the spline-based path planning method, which accommodates to AUV's practically limited perceiving capability.
- An NMPC-based tracking controller is designed to track the planned reference path. Sufficient conditions on closed-loop stability are derived.
- A unified RHO scheme, seamlessly integrating the path planning and the MPC tracking control, is proposed for the AUV application.

3.1.2 Chapter Organization

The remainder of this chapter is organized as follows: In Sect. 3.2, the problem statement is presented. In Sect. 3.3, we develop a receding horizon path planning method for the AUV. Section 3.4 provides the formulation of the tracking control problem into the NMPC scheme as well as the detailed controller design. In Sect. 3.5, the overall motion control algorithm which combines the NMPC tracking control and the path planning is depicted. Simulation results are demonstrated in Sect. 3.6. Section 3.7 provides the concluding remarks.

The notation adopted in this chapter is explained as follows: The symbol $\|\cdot\|$ refers to the Euclidean norm for vectors and the induced 2-norm for matrices. For a function $f(x)$, the time derivative is denoted by $\dot{f}(x)$ while the derivative with respect to x is denoted by $f'(x)$. The diagonal operation is abbreviated by $\text{diag}(\cdot)$.

3.2 Problem Statement

The combined problem of path planning and tracking control arises in many AUV applications. For example, in the marine source seeking applications (Brinón-Arranz et al. 2014), or deep-sea archeological exploration applications (Bingham et al. 2010), the workspace is constrained or cluttered with obstacles. As illustrated in Fig. 3.1, the workspace of AUVs is confined by two polygonal chains $c_1(x)$ and $c_2(x)$ which reflect the topography of the seafloor or the structure of an underwater shipwreck. Inevitably, path planning needs to be taken into account for safety reasons.

Definitions of reference paths may be diverse for various applications. A reference path should be continuous, feasible, and maybe with some performance index optimized, e.g., the way-point straight path for marine vessels (Fossen 2002). A more advanced path planning algorithm requires a higher order of smoothness, i.e., the resultant path is continuous concerning its derivatives. Note that the second-order derivative of the path is proportional to the vehicle acceleration, a reference path with a continuous second-order derivative is particularly pursued.

Since the AUV sway velocity is considerably smaller than its surge counterpart, the following nonholonomic kinematic equations well approximate the vehicle motion (Fossen 2002) at the velocity level:

$$\begin{aligned} \dot{x} &= u\cos\psi - v\sin\psi \approx u_0\cos\psi, \\ \dot{y} &= u\sin\psi + v\cos\psi \approx u_0\sin\psi, \\ \dot{\psi} &= r, \end{aligned} \tag{3.1}$$

where $[x, y, \psi]^{\mathrm{T}}$ denote the position and orientation; $[u, v, r]^{\mathrm{T}}$ stand for the surge, sway and yaw velocities while u_0 denotes the resultant tangential velocity of the vehicle.

To precisely track a desired parameterized path $p(s) = [x_d(s), y_d(s)]^{\mathrm{T}}$, the required open-loop tangential velocity u_{oc} and yaw rate r_{oc} can be calculated for the kinematic model (3.1) in an explicit manner, provided that there is no disturbance or initial error

$$u_{\mathrm{oc}}(t) = \sqrt{x_d'^2 + y_d'^2}\dot{s}, \tag{3.2a}$$

$$r_{\mathrm{oc}}(t) = \frac{x_d' y_d'' - y_d' x_d''}{x_d'^2 + y_d'^2}\dot{s} = u_{\mathrm{oc}}(t)\kappa(s), \tag{3.2b}$$

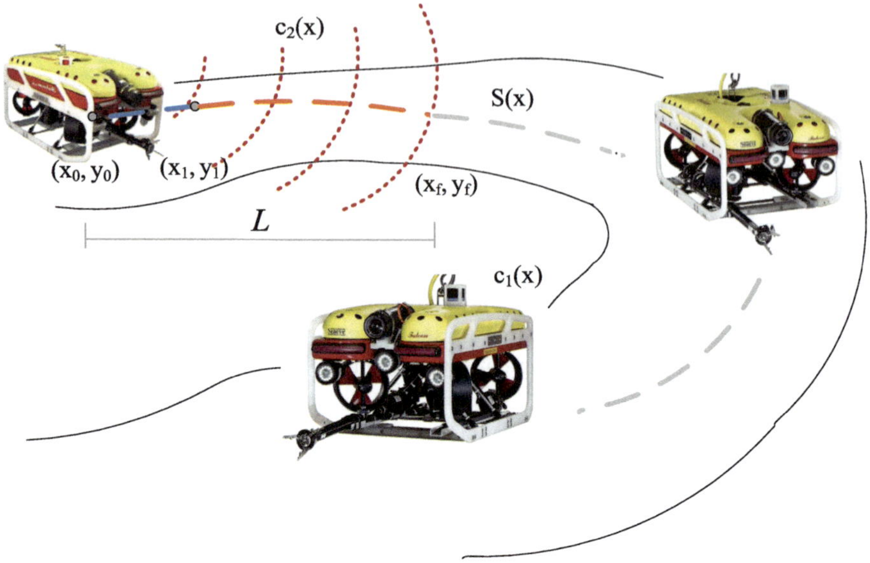

Fig. 3.1 Illustration of the combined AUV motion control problem. Photo courtesy of Saab Seaeye Ltd. ©2022 Saab Seaeye Ltd, reproduced with permission from (Saab Seaeye Ltd. 2022)

where $\kappa(s)$ represents the curvature along the path. From (3.2) it is readily seen that the smaller curvature will make the steering easier. In this sense, a reference path with minimum curvature is practically desirable for AUVs.

However, due to the quick decaying rate of acoustic signals in the water, the effective sensing range is practically short (supposed to be L meters in Fig. 3.1). The reference path needs to be dynamically generated as the vehicle moves forward, i.e., the path planning and tracking control are inherently coupled. Therefore, the AUV control task contains the following two aspects:

(1) *Dynamic path planning:* The AUV can dynamically plan a minimum curvature reference path based on locally updated information.
(2) *Precise tracking control:* The AUV can precisely converge and track the planned reference path.

3.3 Minimum Curvature Receding Horizon Path Planning

Inspired by (Jung and Tsiotras 2013) and (Berglund et al. 2010), we take advantage of spline functions as the path template. Spline functions have many desirable properties that make them suitable for path planning problems in the engineering practice (De Boor and De Boor 1978). A spline can be conveniently selected to provide enough order of smoothness and is parameterized to facilitate the reference path computation.

A spline function represented in b-form is defined as follows:

$$S(x) = \sum_{i=1}^{n} \alpha_i S_{i,k}(x), \tag{3.3}$$

where n is the number of knots, α_i are the control parameters and $S_{i,k}(x)$ are kth-order basis splines defined recursively,

$$S_{i,1}(x) = \begin{cases} 1, & x_i \le x < x_{i+1}, \\ 0, & \text{otherwise}, \end{cases}$$

and

$$S_{i,k}(x) = \frac{(x - x_i)S_{i,k-1}(x)}{x_{i+k-1} - x_i} + \frac{(x_{i+k} - x)S_{i+1,k-1}(x)}{x_{i+k} - x_{i+1}}.$$

For the spline function $S(x)$, the measure of smoothness is chosen as (3.4), which is the objective function in the reference path planning calculation.

$$F(x) = \int_{s(x_0)}^{s(x_f)} \kappa'(s)^2 \mathrm{d}s. \tag{3.4}$$

Considering the relationship between the arc-length s and the Cartesian coordinate x, the objective function in (3.4) is equivalent to

$$F(x) = \int_{x_0}^{x_f} \frac{\kappa'(x)^2}{\sqrt{1 + S'(x)^2}} \mathrm{d}x, \tag{3.5}$$

where $\kappa(x)$ represents the curvature of the spline

$$\kappa(x) = \frac{S''(x)}{(1 + S'(x)^2)^{3/2}}.$$

For simplicity, we assume that the yaw plane environment is restricted by two polygonal chains $c_1(x)$ and $c_2(x)$ as shown in Fig. 3.1. Then the path planning problem is formulated as the following optimization problem:

Problem 3.1 *Given the restriction polygonal chains $c_1(x) \le c_2(x)$, with proper numbers of knots n, order o and end multiplicity l, compute the spline function $S(\alpha, x)$ defined on $\Omega = [x_0, x_f]$ such that*

$$\begin{aligned} &\min_{\alpha} \; F(\alpha, x) \\ &\text{s.t.} \;\; c_1(x) \le S(\alpha, x) \le c_2(x), \end{aligned}$$

where $\alpha = [\alpha_1, \ldots, \alpha_n]^T$. Here, we denote the spline as $S(\alpha, x)$ and objective function as $F(\alpha, x)$ to explicitly point out the dependence on the control parameters which are the decision variables in the optimization. In other places, when the spline is already determined, we simply use the notation $S(x)$ as per established convention (De Boor and De Boor 1978). Also we notice that the order of the spline template should be at least three to guarantee the continuity of the acceleration.

Remark 3.1 It is worth noting that the number of control parameters has an explicit relation to the number of knots; thus, the selection of knots appears a significant issue. Extensive studies on knot placement can be found in (Li et al. 2005). For the AUV application, the larger number of knots makes the description of the environment more accurate, but it also makes the calculation of reference path $S(x)$ more time-consuming. In view of the limited computational resource and memory size on the chip, the number of knots is usually required to be as small as possible without losing the main characteristics of the environment. A possible way is to choose the local extrema of the environment and connect these extrema sequentially. A safe margin concept (Berglund et al. 2010) can be used as well.

Remark 3.2 Another issue regarding computational complexity is the inequality constraints $c_1(x) \leq S(\alpha, x) \leq c_2(x)$. The constraints are enforced numerically, i.e., we first sample these functions and then impose the inequality constraints at discrete points. Therefore, to increase the number of samples is to enlarge the size of the optimization problem. Fortunately, b-splines form a stable basis for splines, and Runge's phenomenon in which oscillation may occur between knots can be avoided. In practice, a small number of samples are usually sufficient.

Remark 3.3 Although we simplify the yaw plane environment as the area bounded by two polygonal chains, a more complicated environment can be tackled by transforming it into constraints of the optimization problem (or equivalently feasible regions) for the NLP algorithm. The NLP algorithms such as the SQP method and interior-point method (Antoniou and Lu 2007) usually require a connected feasible region. However, a complicated environment may be characterized as several disjoint feasible subregions. This will result in more than one time NLP to calculate the reference path.

Because the sensing range is limitedly short in practice, a global optimization cannot be performed for the AUV path planning. Instead, the receding horizon scheme can help to approximate the minimum curvature reference path. This idea can be explained through the following example: Suppose the sensing range is L-meter, at a specific position, a local reference path defined on $[x_0, x_f]$ (illustrated by the blue plus orange line segments in Fig. 3.1) is generated based on the current L-meter data; when the AUV moves one meter forward (illustrated by the blue line segment) to x_1, a new reference path, between the new start point $\bar{x}_0 = x_1$ and new end point $\bar{x}_f$, will be generated based on the new L-meter measurement data. In this case, the continuity between the neighbor splines has to be considered. Therefore, we impose the continuity conditions as constraints in the optimization formulation.

Problem 3.2 *Given measured environment information $c_1(x) \leq c_2(x)$ only defined on $\Omega = [x_0, x_f]$ and the last calculated reference path $S_0(x)$, with proper numbers of knots n, order o and end multiplicity l, compute the spline function $S(\alpha, x)$ defined on Ω such that*

$$
\begin{aligned}
\min_{\alpha} \;& F(\alpha, x) \\
\text{s.t.} \;& c_1(x) \leq S(\alpha, x) \leq c_2(x) \\
& \|S(\alpha, x_0) - S_0(x_0)\| \leq \epsilon_1, \\
& \|S'(\alpha, x_0) - S_0'(x_0)\| \leq \epsilon_2, \\
& \|S''(\alpha, x_0) - S_0''(x_0)\| \leq \epsilon_3,
\end{aligned}
$$

where ϵ_i is the tolerance for the discontinuity. We notice that the start point of the current spline x_0 is coincident with the second knot of the previous calculated spline.

3.4 Nonlinear Model Predictive Tracking Control

In this section, we propose the nonlinear model predictive tracking control method such that the AUV can precisely track the reference path generated in Sect. 3.3.

The AUV model studied for the tracking control is the Falcon dynamic model which we have discussed in Chap. 2

$$
\dot{\mathbf{x}} = \begin{bmatrix} \mathbf{R}(\psi)\mathbf{v} \\ \mathbf{M}^{-1}(\mathbf{Bu} - \mathbf{C}(\mathbf{v})\mathbf{v} - \mathbf{D}(\mathbf{v})\mathbf{v} - \mathbf{g}(\boldsymbol{\eta})) \end{bmatrix} = \mathbf{f}(\mathbf{x}, \mathbf{u}), \tag{3.6}
$$

where the state vector $\mathbf{x} = [x, y, \psi, u, v, r]^{\mathrm{T}}$ consists of the pose and velocity of the vehicle, and the control vector $\mathbf{u} = [u_1, u_2, u_3, u_4]^{\mathrm{T}}$ includes the forces generated by the four thrusters. The detailed expression can be found in (2.32) and (2.34).

From a control point of view, the trajectory-tracking control is challenging due to the nonlinearity and coupled system dynamics (3.6). Since the vehicle surge velocity does not keep constant for the trajectory-tracking task, the assumptions for model linearization do not hold, and linear control methods are inappropriate to apply. In this regard, NMPC, which is capable of dealing with complex nonlinearity of the dynamics, seems an attractive and practical option.

The reference path $S(x)$ can be viewed as the state trajectory $\boldsymbol{\eta}_d = [x_d, y_d, \psi_d]^{\mathrm{T}}$ generated by a reference system which owns the same kinematic equations of motion as the real vehicle's

$$
\begin{aligned}
\dot{x}_d &= u_d \cos\psi_d - v_d \sin\psi_d, \\
\dot{y}_d &= u_d \sin\psi_d + v_d \cos\psi_d, \\
\dot{\psi}_d &= r_d.
\end{aligned} \tag{3.7}
$$

Decompose the kinematic state error in the AUV parallel reference frame (see Fig. 3.2)

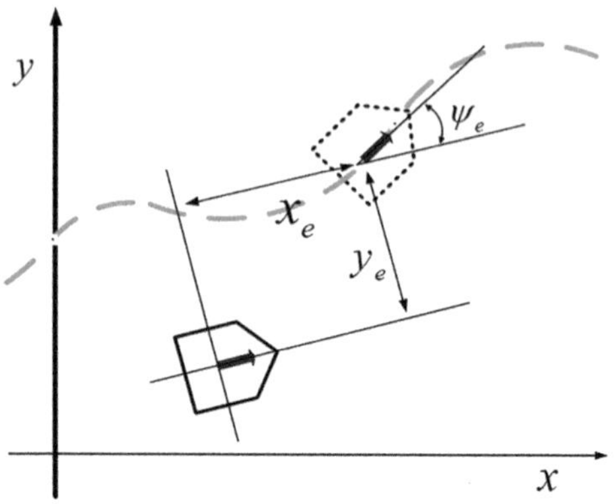

Fig. 3.2 Illustration of the kinematic error in the AUV parallel frame

$$\boldsymbol{\eta}_e = \begin{bmatrix} x_e \\ y_e \\ \psi_e \end{bmatrix} = \begin{bmatrix} \cos\psi & \sin\psi & 0 \\ -\sin\psi & \cos\psi & 0 \\ 0 & 0 & 1 \end{bmatrix} \begin{bmatrix} x_d - x \\ y_d - y \\ \psi_d - \psi \end{bmatrix}. \tag{3.8}$$

We further partition the velocity $\mathbf{v} = \mathbf{v}_f + \mathbf{v}_e$, where $\mathbf{v}_f = [u_d \cos\psi_e - v_d \sin\psi_e, u_d \sin\psi_e + v_d \cos\psi_e, r_d]^{\mathrm{T}}$ can be regarded as a feedforward control action, and $\mathbf{v}_e = [u_e, v_e, r_e]^{\mathrm{T}}$ represents the feedback control action. Differentiating both sides of (3.8) and substituting (2.34) (3.7), we derive the kinematic error equations

$$\begin{aligned} \dot{x}_e &= y_e r_d - u_e + y_e r_e, \\ \dot{y}_e &= -x_e r_d - v_e - x_e r_e, \\ \dot{\psi}_e &= -r_e. \end{aligned} \tag{3.9}$$

Analogously, we view $\mathbf{v}_d = [u_d, v_d, r_d]^{\mathrm{T}}$ as the state trajectory of the reference system that is with the same dynamic equations of motion

$$\begin{aligned} \dot{u}_d &= \frac{M_{\dot{v}}}{M_{\dot{u}}} v_d r_d - \frac{X_u}{M_{\dot{u}}} u_d - \frac{D_u}{M_{\dot{u}}} u_d |u_d| + \frac{F_{ud}}{M_{\dot{u}}}, \\ \dot{v}_d &= -\frac{M_{\dot{u}}}{M_{\dot{v}}} u_d r_d - \frac{Y_v}{M_{\dot{v}}} v_d - \frac{D_v}{M_{\dot{v}}} v_d |v_d| + \frac{F_{vd}}{M_{\dot{v}}}, \\ \dot{r}_d &= \frac{M_{\dot{u}} - M_{\dot{v}}}{M_{\dot{r}}} u_d v_d - \frac{N_r}{M_{\dot{r}}} r_d - \frac{D_r}{M_{\dot{r}}} r_d |r_d| + \frac{F_{rd}}{M_{\dot{r}}}. \end{aligned} \tag{3.10}$$

Differentiating both sides of $\mathbf{v}_e = \mathbf{v} - \mathbf{v}_f$ yields

$$\begin{aligned} \dot{u}_e &= \dot{u} - \dot{u}_d \cos\psi_e - u_d \sin\psi_e r_e + \dot{v}_d \sin\psi_e - v_d \cos\psi_e r_e, \\ \dot{v}_e &= \dot{v} - \dot{u}_d \sin\psi_e + u_d \cos\psi_e r_e - \dot{v}_d \cos\psi_e - v_d \sin\psi_e r_e, \\ \dot{r}_e &= \dot{r} - \dot{r}_d. \end{aligned} \tag{3.11}$$

Substituting (2.32) into (3.11), we have

$$\begin{aligned}\dot{u}_e = &\tfrac{M_{\dot{v}}}{M_{\dot{u}}}(v_e + u_d \sin\psi_e + v_d \cos\psi_e)(r_e + r_d) - \tfrac{X_u}{M_{\dot{u}}}(u_e + u_d \cos\psi_e \\ &- v_d \sin\psi_e) - \tfrac{D_u}{M_{\dot{u}}}(u_e + u_d \cos\psi_e - v_d \sin\psi_e)|u_e + u_d \cos\psi_e - v_d \sin\psi_e| \\ &- \dot{u}_d \cos\psi_e - u_d \sin\psi_e r_e + \dot{v}_d \sin\psi_e - v_d \cos\psi_e r_e + \tfrac{F_u}{M_{\dot{u}}},\end{aligned} \tag{3.12a}$$

$$\begin{aligned}\dot{v}_e = &-\tfrac{M_{\dot{u}}}{M_{\dot{v}}}(u_e + u_d \cos\psi_e - v_d \sin\psi_e)(r_e + r_d) - \tfrac{Y_v}{M_{\dot{v}}}(v_e + u_d \sin\psi_e \\ &+ v_d \cos\psi_e) - \tfrac{D_v}{M_{\dot{v}}}(v_e + u_d \sin\psi_e + v_d \cos\psi_e)|v_e + u_d \sin\psi_e + v_d \cos\psi_e| \\ &- \dot{u}_d \sin\psi_e + u_d \cos\psi_e r_e - \dot{v}_d \cos\psi_e - v_d \sin\psi_e r_e + \tfrac{F_v}{M_{\dot{v}}},\end{aligned} \tag{3.12b}$$

$$\begin{aligned}\dot{r}_e = &\tfrac{M_{\dot{u}} - M_{\dot{v}}}{M_{\dot{r}}}(u_e + u_d \cos\psi_e - v_d \sin\psi_e)(v_e + u_d \sin\psi_e + v_d \cos\psi_e) \\ &- \tfrac{N_r}{M_{\dot{r}}}(r_e + r_d) - \tfrac{D_r}{M_{\dot{r}}}(r_e + r_d)|r_e + r_d| - \dot{r}_d + \tfrac{F_r}{M_{\dot{r}}}.\end{aligned} \tag{3.12c}$$

Design F_u, F_v, and F_r as follows:

$$\begin{aligned}F_u = &-M_{\dot{v}}(v_e + u_d \sin\psi_e + v_d \cos\psi_e)(r_e + r_d) + X_u(u_e + u_d \cos\psi_e \\ &- v_d \sin\psi_e) + D_u(u_e + u_d \cos\psi_e - v_d \sin\psi_e)|u_e + u_d \cos\psi_e - v_d \sin\psi_e| \\ &+ M_{\dot{u}}(\dot{u}_d \cos\psi_e + u_d \sin\psi_e r_e - \dot{v}_d \sin\psi_e + v_d \cos\psi_e r_e) + M_{\dot{u}}\tau_u,\end{aligned} \tag{3.13a}$$

$$\begin{aligned}F_v = &M_{\dot{u}}(u_e + u_d \cos\psi_e - v_d \sin\psi_e)(r_e + r_d) + Y_v(v_e + u_d \sin\psi_e \\ &+ v_d \cos\psi_e) + D_v(v_e + u_d \sin\psi_e + v_d \cos\psi_e)|v_e + u_d \sin\psi_e + v_d \cos\psi_e| \\ &+ M_{\dot{v}}(\dot{u}_d \sin\psi_e - u_d \cos\psi_e r_e + \dot{v}_d \cos\psi_e + v_d \sin\psi_e r_e) + M_{\dot{v}}\tau_v,\end{aligned} \tag{3.13b}$$

$$\begin{aligned}F_r = &(M_{\dot{v}} - M_{\dot{u}})(u_e + u_d \cos\psi_e - v_d \sin\psi_e)(v_e + u_d \sin\psi_e \\ &+ v_d \cos\psi_e) + N_r(r_e + r_d) + D_r(r_e + r_d)|r_e + r_d| + M_{\dot{r}}\dot{r}_d + M_{\dot{r}}\tau_r.\end{aligned} \tag{3.13c}$$

By defining $\boldsymbol{\tau}_e = [\tau_u, \tau_v, \tau_r]^{\mathrm{T}}$ and substituting (3.8) into (3.12), together with (3.9) we have the following error dynamics for the AUV tracking control:

$$\dot{\mathbf{x}}_e = \begin{bmatrix} \dot{x}_e \\ \dot{y}_e \\ \dot{\psi}_e \\ \dot{u}_e \\ \dot{v}_e \\ \dot{r}_e \end{bmatrix} = \begin{bmatrix} y_e r_d - u_e + y_e r_e \\ -x_e r_d - v_e - x_e r_e \\ -r_e \\ \tau_u \\ \tau_v \\ \tau_r \end{bmatrix} = \mathbf{f}_e(\mathbf{x}_e, \boldsymbol{\tau}_e). \tag{3.14}$$

Obviously, $(\mathbf{0}, \mathbf{0})$ is an equilibrium point of (3.14). Comparing (3.10) and (3.8) at $(\mathbf{0}, \mathbf{0})$, we find that $F_u = F_{ud}$, $F_v = F_{vd}$, and $F_r = F_{rd}$, which means that the tracking of the reference trajectory is equivalent to the stabilizing the error system (3.14) to the equilibrium point.

The NMPC technique is used to stabilize the error dynamics. The stabilization is realized through the minimization of the cost function:

$$J(t, \mathbf{x}_e(t), \boldsymbol{\tau}_e(t)) = g(\mathbf{x}_e(t+T)) + \int_t^{t+T} \ell(\mathbf{x}_e(s), \boldsymbol{\tau}_e(s))\mathrm{d}s, \tag{3.15}$$

where the initial time is viewed as 0, T is the prediction horizon, $g(\cdot)$ is the terminal state penalty satisfying $g(\mathbf{0}) = 0$ and $g(\mathbf{x}_e) > 0$ for any $\mathbf{x}_e \neq \mathbf{0}$; $\ell(\mathbf{x}_e, \boldsymbol{\tau}_e) = \mathbf{x}_e^{\mathrm{T}}\mathbf{Q}\mathbf{x}_e + \boldsymbol{\tau}_e^{\mathrm{T}}\mathbf{R}\boldsymbol{\tau}_e$ is the stage cost with the positive definite matrices $\mathbf{Q}$ and $\mathbf{R}$. Then the optimal control problem to be solved at each sampling time instant is formulated as follows:

Problem 3.3 *Given the current error state $\mathbf{x}_e(t)$, the weighting matrices $\mathbf{Q}$ and $\mathbf{R}$, compute the optimal control force $\boldsymbol{\tau}_e(t)$ by solving the following optimization problem:*

$$\begin{aligned} \min_{\boldsymbol{\tau}_e} \ & J(t, \mathbf{x}_e(t), \boldsymbol{\tau}_e(t)) \\ \text{s.t.} \ & \dot{\mathbf{x}}_e(s) = \mathbf{f}_e(\mathbf{x}_e(s), \boldsymbol{\tau}_e(s)), \\ & \mathbf{x}_e(s) \in X_e, \\ & \boldsymbol{\tau}_e(s) \in T_e, \\ & \mathbf{x}_e(t+T) \in X_f \subset X_e, \end{aligned} \tag{3.16}$$

where $s \in [t, t+T)$, X_e is a closed set representing the constraints on the error state, T_e is a compact set containing all the allowed control inputs, and X_f is the terminal constraint set.

The standard NMPC algorithm for the AUV tracking control problem can be briefly described as follows:

- The optimization problem (3.16) is solved with the current error state $\mathbf{x}_e(t)$ as the initial condition. Let $\bar{\boldsymbol{\tau}}_e(t)$ denote the solution.
- The system uses $\bar{\boldsymbol{\tau}}_e(t)$ to calculate $\boldsymbol{\tau}$ using (3.13) for only one sampling period: $\boldsymbol{\tau}_e(t) = \bar{\boldsymbol{\tau}}_e(t)$ for $[t, t+\delta]$.
- At time $t+\delta$, the optimization problem (3.16) is solved again using the latest measured state data $\mathbf{x}_e(t+\delta)$.

The above procedure will be repeated until accomplishing the AUV tracking task. However, it is well known that optimality does not guarantee the closed-loop stability due to the finite time prediction horizon. To solve this problem, we need to appropriately design the local controller $\boldsymbol{\tau}_e^L$, the terminal state penalty $g(\cdot)$, and the terminal constraint X_f (Mayne et al. 2000).

Theorem 3.1 *The closed-loop system (3.14) controlled by the NMPC algorithm is asymptotically stable if the following conditions are satisfied:*

C1: $\mathbf{0} \in X_f$ *and* $\mathbf{0} \in T_e$*;*

C2: There exists a local controller $\boldsymbol{\tau}_e^L(t)$ *satisfying*

(a) $\boldsymbol{\tau}_e^L(t) \in T_e$, *for all* $\mathbf{x}_e \in X_f$,

(b) $\mathbf{f}_e(\mathbf{x}_e, \boldsymbol{\tau}_e^L(t)) \in X_f$, *for all* $\mathbf{x}_e \in X_f$,

(c) $\dot{g}(\mathbf{x}_e) + \ell(\mathbf{x}_e, \boldsymbol{\tau}_e^L(t)) \leq 0$, *for all* $\mathbf{x}_e \in X_f$*;*

C3: Let $\bar{\boldsymbol{\tau}}_e(t|t_0)$ *denote the solution of the optimization problem (3.16) at time* t_0*, and the initial guess for the optimization at the next sampling time instant* $t_0 + \delta$ *is chosen as*

$$\hat{\boldsymbol{\tau}}_e(t|t_0 + \delta) = \begin{cases} \bar{\boldsymbol{\tau}}_e(t|t_0), & t_0 + \delta \leq t \leq t_0 + T, \\ \boldsymbol{\tau}_e^L(t), & t_0 + T \leq t \leq t_0 + T + \delta. \end{cases} \tag{3.17}$$

Proof To prove the stability of the NMPC algorithm, we need to find a Lyapunov function such that this function is nonincreasing along the trajectory of (3.14) controlled by the NMPC algorithm. We take the optimal value function V of the performance index (3.15) as a Lyapunov function candidate.

In the following, the value function is denoted as $V(t, \mathbf{x}_e(t_0))$ implying that the optimal value function depends on the initial error; the state trajectory determined by $\bar{\boldsymbol{\tau}}_e(t)$ is denoted by $\bar{\mathbf{x}}_e(t)$; the state trajectory determined by the initial guess (3.17) is $\hat{\mathbf{x}}_e(t)$ and the corresponding cost is $\hat{V}(t, \mathbf{x}_e)$.

(i) For $0 \leq t_1 \leq t_2 < \delta$, V is nonincreasing since

$$\begin{aligned} V(t_2, \mathbf{x}_e(t_0)) &= V(t_1, \mathbf{x}_e(t_0)) - \int_{t_1}^{t_2} \ell(\mathbf{x}_e(s), \boldsymbol{\tau}_e(s))\mathrm{d}s \\ &\leq V(t_1, \mathbf{x}_e(t_0)). \end{aligned}$$

(ii) For $0 \leq t_1 < \delta$, and $\delta \leq t_2 < 2\delta$, we have

$$\begin{aligned} & V(t_2, \mathbf{x}_e(t_0 + \delta)) - V(t_1, \mathbf{x}_e(t_0)) \\ &= V(t_2, \bar{\mathbf{x}}_e(t_0 + \delta)) - V(t_1, \mathbf{x}_e(t_0)) \\ &\leq \hat{V}(t_2, \bar{\mathbf{x}}_e(t_0 + \delta)) - V(t_1, \mathbf{x}_e(t_0)) \\ &= g(\hat{\mathbf{x}}_e(T + \delta)) - g(\bar{\mathbf{x}}_e(T)) + \int_{t_2}^{T+\delta} \ell(\hat{\mathbf{x}}_e(s), \hat{\boldsymbol{\tau}}_e(s))\mathbf{ds} - \int_{t_1}^{T} \ell(\bar{\mathbf{x}}_e(s), \bar{\boldsymbol{\tau}}_e(s))\mathrm{d}s. \end{aligned}$$

Observing (3.17), we notice that for $s \in [t_2, T]$, $\hat{\mathbf{x}}_e(s) = \bar{\mathbf{x}}_e(s)$ and $\hat{\boldsymbol{\tau}}_e(s) = \bar{\boldsymbol{\tau}}_e(s)$, thus we have

$$\begin{aligned} & V(t_2, \mathbf{x}_e(t_0 + \delta)) - V(t_1, \mathbf{x}_e(t_0)) \\ &\leq g(\hat{\mathbf{x}}_e(T + \delta)) - g(\hat{\mathbf{x}}_e(T)) + \int_{T}^{T+\delta} \ell(\hat{\mathbf{x}}_e(s), \hat{\boldsymbol{\tau}}_e(s))\mathrm{d}s - \int_{t_1}^{t_2} \ell(\bar{\mathbf{x}}_e(s), \bar{\boldsymbol{\tau}}_e(s))\mathrm{d}s. \end{aligned}$$

From $C2$-(c), we have $\dot{g}(\hat{\mathbf{x}}_e) + \ell(\hat{\mathbf{x}}_e, \hat{\boldsymbol{\tau}}_e) \leq 0$, then integrating both sides from T to $T + \delta$

$$g(\hat{\mathbf{x}}_e(T + \delta)) - g(\hat{\mathbf{x}}_e(T)) + \int_T^{T+\delta} \ell(\hat{\mathbf{x}}_e(s), \hat{\boldsymbol{\tau}}_e(s))\mathrm{d}s \leq 0.$$

Therefore, we have $V(t_2, \mathbf{x}_e(t_0 + \delta)) \leq V(t_1, \mathbf{x}_e(t_0))$.

(iii) For $k_1\delta \leq t_1 < (k_1 + 1)\delta$ and $k_2\delta \leq t_2 < (k_2 + 1)\delta$, with $k_1 < k_2$, by (ii) the following inequality holds:

$$V(t_2) \leq V(k_2\delta) \cdots \leq V((k_1 + 1)\delta) \leq V(t_1).$$

which implies

$$\begin{aligned} V(t, \mathbf{x}_e(t)) - V(t_0, \mathbf{x}_e(t_0)) &\leq -\int_{t_0}^{t} \ell(\mathbf{x}_e(s), \boldsymbol{\tau}_e(s))\mathrm{d}s \\ &= -\int_{t_0}^{t} \mathbf{x}_e^{\mathrm{T}}\mathbf{Q}\mathbf{x}_e + \boldsymbol{\tau}_e^{\mathrm{T}}\mathbf{R}\boldsymbol{\tau}_e \mathrm{d}s \leq -\int_{t_0}^{t} \mathbf{x}_e^{\mathrm{T}}\mathbf{Q}\mathbf{x}_e \mathrm{d}s. \end{aligned}$$

By construction, we always have $V \geq 0$. Then we can say V is nonincreasing and lower bounded by zero. According to Barbalat's lemma, we have $\mathbf{x}_e^{\mathrm{T}}\mathbf{Q}\mathbf{x}_e \to 0$ as $t \to \infty$. Since $\mathbf{Q}$ is positive definite, the system (3.14) will be steered to the equilibrium point $(\mathbf{0}, \mathbf{0})$ asymptotically by the NMPC algorithm, i.e., $\mathbf{x}_e \to 0$ as $t \to \infty$.

Now, we design the NMPC tracking controller based on Theorem 3.1. First, we construct the initial guess using the way in $C3$. Second, we select the local controller to be a linear state feedback $\boldsymbol{\tau}_e^L(t) = \mathrm{K}\mathbf{x}_e$. Third, we set the terminal state penalty $g(\mathbf{x}_e) \triangleq \frac{1}{2}\mathbf{x}_e^{\mathrm{T}}\mathbf{x}_e$. Specifically, let $\mathrm{K} \triangleq [\mathrm{K}_1\ \mathrm{K}_2]$ where $\mathrm{K}_1 = \mathrm{diag}(\alpha_1, \beta_1, \gamma_1)$ and $\mathrm{K}_2 = -\mathrm{diag}(\alpha_2, \beta_2, \gamma_2)$ with $\alpha_i > 0, \beta_i > 0, \gamma_i > 0$; $\mathbf{Q} = \mathrm{diag}(\mathrm{q}_{11}, \mathrm{q}_{22}, \mathrm{q}_{33}, \mathrm{q}_{44}, \mathrm{q}_{55}, \mathrm{q}_{66})$ and $\mathbf{R} = \mathrm{diag}(\mathrm{r}_{11}, \mathrm{r}_{22}, \mathrm{r}_{33})$ with $q_{ii} > 0$, $r_{ii} \geq 0$, then we calculate the following:

$$\begin{aligned} \dot{g}(\mathbf{x}_e) &= \dot{x}_e x_e + \dot{y}_e y_e + \dot{\psi}_e \psi_e + \dot{u}_e u_e + \dot{v}_e v_e + \dot{r}_e r_e \\ &= (y_e r_d - u_e + y_e r_e)x_e + (-x_e r_d - v_e - x_e r_e)y_e \\ &\quad -\psi_e r_e + u_e \tau_u^L + v_e \tau_v^L + r_e \tau_r^L \\ &= -x_e u_e - y_e v_e - \psi_e r_e + u_e \tau_u^L + v_e \tau_v^L + r_e \tau_r^L, \end{aligned}$$

and

$$\begin{aligned} \ell(\mathbf{x}_e, \boldsymbol{\tau}_e^L(t)) = {} & q_{11}x_e^2 + q_{22}y_e^2 + q_{33}\psi_e^2 + q_{44}u_e^2 \\ & + q_{55}v_e^2 + q_{66}r_e^2 + r_{11}(\tau_u^L)^2 + r_{22}(\tau_v^L)^2 + r_{33}(\tau_r^L)^2. \end{aligned}$$

Substituting $\boldsymbol{\tau}_e^L(t) = \mathrm{K}\mathbf{x}_e$, we have

$$
\begin{aligned}
&\dot{g}(\mathbf{x}_e) + \ell(\mathbf{x}_e, \boldsymbol{\tau}_e^L(t)) \\
&= -x_e u_e - v_e y_e - \psi_e r_e + \alpha_1 x_e u_e - \alpha_2 u_e^2 + \beta_1 y_e v_e \\
&\quad -\beta_2 v_e^2 + \gamma_1 \psi_e r_e - \gamma_2 r_e^2 + q_{11} x_e^2 + q_{22} y_e^2 + q_{33} \psi_e^2 \\
&\quad +q_{44} u_e^2 + q_{55} v_e^2 + q_{66} r_e^2 + r_{11}(\alpha_1 x_e - \alpha_2 u_e)^2 \\
&\quad +r_{22}(\beta_1 y_e - \beta_2 v_e)^2 + r_{33}(\gamma_1 \psi_e - \gamma_2 r_e)^2 \\
&= (q_{44} - \alpha_2 + r_{11}\alpha_2^2) u_e^2 + (q_{55} - \beta_2 + r_{22}\beta_2^2) v_e^2 \\
&\quad +(q_{66} - \gamma_2 + r_{33}\gamma_2^2) r_e^2 + (q_{11} + r_{11}\alpha_1^2) x_e^2 \\
&\quad +(\alpha_1 - 1 - 2\alpha_1\alpha_2 r_{11}) x_e u_e + (q_{22} + r_{22}\beta_1^2) y_e^2 \\
&\quad +(\beta_1 - 1 - 2\beta_1\beta_2 r_{22}) y_e v_e + (q_{33} + r_{33}\gamma_1^2) \psi_e^2 \\
&\quad +(\gamma_1 - 1 - 2\gamma_1\gamma_2 r_{33}) \psi_e r_e .
\end{aligned}
$$

We select these parameters which satisfy the following inequalities:

$$
\begin{aligned}
q_{11} + r_{11}\alpha_1^2 + \alpha_1 - 1 - 2\alpha_1\alpha_2 r_{11} &\leq 0, \\
q_{22} + r_{22}\beta_1^2 + \beta_1 - 1 - 2\beta_1\beta_2 r_{22} &\leq 0, \\
q_{33} + r_{33}\gamma_1^2 + \gamma_1 - 1 - 2\gamma_1\gamma_2 r_{33} &\leq 0, \\
q_{44} - \alpha_2 + r_{11}\alpha_2^2 &\leq 0, \\
q_{55} - \beta_2 + r_{22}\beta_2^2 &\leq 0, \\
q_{66} - \gamma_2 + r_{33}\gamma_2^2 &\leq 0,
\end{aligned} \tag{3.18}
$$

and select the terminal constraint to satisfy

$$
\begin{aligned}
|u_e| \geq |x_e|, \;\; x_e u_e \geq 0, \\
|v_e| \geq |y_e|, \;\; y_e v_e \geq 0, \\
|r_e| \geq |\psi_e|, \;\; \psi_e r_e \geq 0.
\end{aligned} \tag{3.19}
$$

We further choose the terminal constraint set to be the sublevel set of $h(\cdot)$

$$
X_{f1} = \{\mathbf{x}_e \in \mathbb{R}^6 \mid \frac{1}{2}\mathbf{x}_e^{\mathrm{T}}\mathbf{x}_e \leq \sigma\}. \tag{3.20}
$$

As a result, the condition $C2$ is satisfied. Define $X_f = X_{f1} \cap X_{f2}$, where

$$
X_{f2} = \{\mathbf{x}_e \in \mathbb{R}^6 \mid (3.19)\}. \tag{3.21}
$$

Here, we assume $\mathbf{0} \in T_e$, then $C1$ can be satisfied in X_f.

Remark 3.4 The inequalities in (3.18) are easy to satisfy. In fact, these parameters can be divided into three independent groups: $(q_{11}, q_{44}, r_{11}, \alpha_1, \alpha_2)$, $(q_{22}, q_{55}, r_{22}, \beta_1, \beta_2)$, and $(q_{33}, q_{66}, r_{33}, \gamma_1, \gamma_2)$. The selection among these groups is not coupled. Then for each group, only two inequality conditions need to be satisfied.

Remark 3.5 Back to (3.7) and (3.10), we notice that the reference system plays a crucial role in the NMPC tracking algorithm. Generally speaking, finding such a reference system is nontrivial, and it might be even more difficult than the stabilization problem itself. Fortunately, for the AUV tracking problem, the reference pair $(\mathbf{x}_d, \boldsymbol{\tau}_d)$

can be derived directly from the reference path $S(x)$. In particular, one possible option can be determined in the following way:

- ☐ Make the calculated spline $S(x)$ a reference trajectory of the kinematic state $\boldsymbol{\eta}_d$ by employing a predetermined timing law:

$$\begin{aligned} x_d &= \rho t, \\ y_d &= S(x_d), \\ \psi_d &= \text{atan2}(\dot{y}_d, \dot{x}_d), \end{aligned}$$

  where $\rho > 0$ and atan2$(\cdot, \cdot)$ is the four-quadrant inverse tangent operator.
- ☐ Choose the reference velocity $\mathbf{v}_d$ to be $u_d = u_{\text{oc}}$, $v_d = 0$ and $r_d = r_{\text{oc}}$.
- ☐ Calculate the reference control input $\boldsymbol{\tau}_d$ through (3.10).

Remark 3.6 For the AUV tracking control, the constraints on control inputs are usually given as follows:

$$F_{u,\min} \le F_u \le F_{u,\max}, \tag{3.22a}$$

$$F_{v,\min} \le F_v \le F_{v,\max}, \tag{3.22b}$$

$$F_{r,\min} \le F_r \le F_{r,\max}. \tag{3.22c}$$

Then we can derive the control error constraints as

$$\frac{F_{u,\min} - F_1}{M_{\dot{u}}} \le \tau_u \le \frac{F_{u,\max} - F_1}{M_{\dot{u}}}, \tag{3.23a}$$

$$\frac{F_{v,\min} - F_2}{M_{\dot{v}}} \le \tau_v \le \frac{F_{v,\max} - F_2}{M_{\dot{v}}}, \tag{3.23b}$$

$$\frac{F_{r,\min} - F_3}{M_{\dot{r}}} \le \tau_r \le \frac{F_{r,\max} - F_3}{M_{\dot{r}}}. \tag{3.23c}$$

Here, $F_1 = F_u - M_{\dot{u}}\tau_u$, $F_2 = F_v - M_{\dot{v}}\tau_v$, and $F_3 = F_r - M_{\dot{r}}\tau_r$ which can be derived from (3.13a)–(3.13c). Although (3.23) is time-varying with the reference system, it is not necessary to explicitly calculate (3.23). The optimization problem (3.16) is defined to facilitate the description of the stability theorem. In the NMPC implementation, we can directly use the polyhedral input constraint (3.22) to calculate $\boldsymbol{\tau}$ but construct the cost function (3.15) using the elaborately defined error state $\mathbf{x}_e$ and control $\boldsymbol{\tau}_e$. Then the equivalent $\boldsymbol{\tau}_e = \boldsymbol{\tau} - F_F$ with $F_F = [F_1, F_2, F_3]^{\mathrm{T}}$.

Notice that the predetermined timing law can be carefully designed (Frezza 1999; Saccon et al. 2011) such that the reference system chosen in Remark 3.5 fits the actual system. Specifically, when the AUV is moving along the reference spline $S(x)$, the reference control input can be expressed in an explicit manner

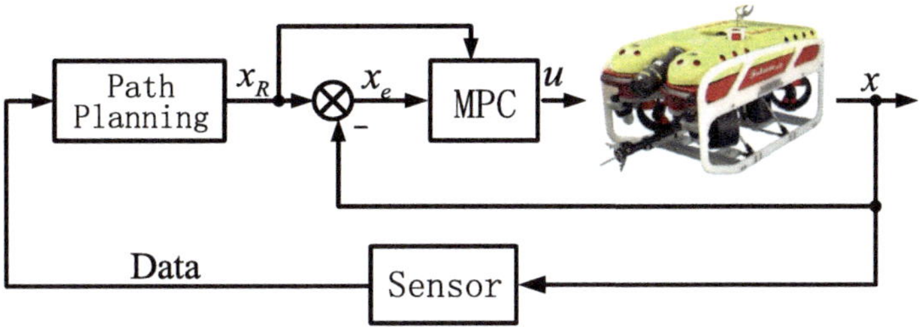

Fig. 3.3 Closed-loop control block diagram. Photo courtesy of Saab Seaeye Ltd. ©2022 Saab Seaeye Ltd, reproduced with permission from (Saab Seaeye Ltd. 2022)

$$F_{ud} = M_{\dot{u}}\dot{u}_d + X_u u_d + D_u u_d|u_d|, \tag{3.24a}$$
$$F_{vd} = M_{\dot{u}} u_d r_d, \tag{3.24b}$$
$$F_{rd} = M_{\dot{r}}\dot{r}_d + N_r r_d + D_r r_d|r_d|, \tag{3.24c}$$

where $\dot{u}_d$ and $\dot{r}_d$ can also be explicitly calculated by

$$\dot{u}_d = S'(x_d)S''(x_d)(1 + S'(x_d)^2)^{-\frac{1}{2}}\dot{x}_d^2 + (1 + S'(x_d)^2)^{\frac{1}{2}}\ddot{x}_d, \tag{3.25a}$$
$$\dot{r}_d = \frac{(1 + S'(x_d)^2)S'''(x_d) - 2S'(x_d)S''(x_d)^2}{(1 + S'(x_d)^2)^2}\dot{x}_d^2 + \frac{S''(x_d)}{(1 + S'(x_d)^2)}\ddot{x}_d. \tag{3.25b}$$

Together with the reference state defined in Remark 3.5, it can be observed that we can always adjust x_d, i.e., the value of ρ to satisfy the input constraints. In the simulations, we use a fixed ρ as a simple example.

In this chapter, the thrust allocation problem is solved by conventional pseudoinverse method, i.e., the calculated generalized generalized thrust forces and moments $\boldsymbol{\tau} = [F_u, F_v, F_r]^{\mathrm{T}}$ will be transformed into real thrusts by $\mathbf{u} = \mathbf{B}^{+}\boldsymbol{\tau}$, where $\mathbf{B}^{+}$ is the Moore–Penrose pseudoinverse.

3.5 Integrated Receding Horizon Path Planning and Tracking Control: Algorithm Description

In view of the same RHO nature, the NMPC tracking control can be conveniently integrated with the proposed path planning method.

Let $\mathcal{D}_i$ denote the measured data for local environment information. The dimension of sampled reference path $S(x)$ between the first and second knot is $\mathcal{M}$. And the prediction horizon $T = \mathcal{N}\delta$ with $\mathcal{N} < \mathcal{M}$. Based on Problems 3.1–3.3, the integrated path planning and tracking control algorithm is summarized in Algorithm 3.1.

In Algorithm 3.1 two things need to be clarified. First, within the **for** iterations, we simply use $\dot{\mathbf{x}} = \mathbf{f}(\mathbf{x}, \mathbf{u})$ to denote the control process. Second, since the prediction

Algorithm 3.1 Integrated Path Planning and Tracking Control Algorithm

$i = 0$.
Let $\mathbf{x}_0$ denote the initial state of the AUV.
Given sensor data $\mathcal{D}_i$, calculate the reference path by solving Problem 3.1. Let $S_i(x)$ be the solution.
procedure
  $k = 0$.
  for $k \leq \mathcal{M} - \mathcal{N}$ **do**
    Given $S_i(x)$ calculate the reference pair $(\mathbf{x}_d, \boldsymbol{\tau}_d)$ according to *Remark* 3.5.
    Solve Problem 3.3 and use (3.8) to calculate $\bar{\boldsymbol{\tau}}(t)$; $\mathbf{u} = \mathbf{B}^+\bar{\boldsymbol{\tau}}$.
    Implement the control: $\dot{\mathbf{x}} = \mathbf{f}(\mathbf{x}, \mathbf{u})$ for $k\delta \leq t \leq (k+1)\delta$.
    $k = k + 1$.
  end for
  $i = i + 1$.
  Collect another set of sensor data $\mathcal{D}_i$.
  Calculate reference path by solving Problem 3.2; let $S_i(x)$ be the solution.
  for do$k \leq \mathcal{M}$
    Given $S_{i-1}(x)$ and $S_i(x)$ calculate $(\mathbf{x}_d, \boldsymbol{\tau}_d)$ according to *Remark* 3.5.
    Solve Problem 3.3 and use (3.8) to calculate $\bar{\boldsymbol{\tau}}(t)$; $\mathbf{u} = \mathbf{B}^+\bar{\boldsymbol{\tau}}$.
    Implement the control: $\dot{\mathbf{x}} = \mathbf{f}(\mathbf{x}, \mathbf{u})$ for $k\delta \leq t \leq (k+1)\delta$.
    Let $\mathbf{x}_f$ denote the system state at time $(k+1)\delta$.
    $k = k + 1$.
  end for
end procedure
$\mathbf{x}_0 = \mathbf{x}_f$, repeat **procedure.**

horizon $\mathcal{N}$ is fixed and the path planning is performed by RHO, when the first segment of the current spline between the first knot and second knot cannot provide enough samples of reference state $\mathbf{x}_d(k)$ from $k = 1$ to $\mathcal{N}$, i.e., $\mathcal{M} - k < \mathcal{N}$, another round of path planning needs to be performed starting from the second knot of the current spline.

The block diagram which represents the integrated control scheme consisting of the path planner and the NMPC tracking controller is depicted in Fig. 3.3. With the aid of block diagram, we can explain the **procedure** in Algorithm 3.1 as follows:

- ☐ The local environment $\mathcal{D}_i$ is detected. The mathematical representation of the workspace (c_1, c_2) is calculated according to current state $\mathbf{x}$ of the AUV, based on which the path planner constructs a problem, i.e., Problem 3.2 to determine the minimum curvature spline path $S_i(x)$.
- ☐ The path $S_i(x)$ is then augmented and viewed as the trajectory of a virtual reference system which provides the reference state $(\mathbf{x}_d, \boldsymbol{\tau}_d)$ for the NMPC tracking controller. A feed-forward channel exists to calculate the control effort F_F corresponding to the compensation of the time-varying portion of the reference path. Based on an elaborately defined error dynamics $(\mathbf{x}_e, \boldsymbol{\tau}_e)$, the receding horizon optimization problem, i.e., Problem 3.3 is constructed and then solved to obtain the tracking control signal $\mathbf{u} = \mathbf{B}^+\bar{\boldsymbol{\tau}}$.
- ☐ As the AUV moves, the newly measured data will be collected and used to generate an updated reference path. Repeat **procedure**.

For the NMPC tracking control, it is worth noting that at each sampling instant, only $\mathcal{N}$ steps of the reference states are used to construct the optimization problem. According to Algorithm 3.1, we have $\mathcal{N} < \mathcal{M}$ and every time (except for the first time) the reference path is planned starting from $\mathcal{N}$ steps ahead of the vehicle position. In this case, the reference path is equivalent to a predefined one although it is actually generated as the vehicle moves forward. The closed-loop stability of the tracking control will not be affected by the integration of the receding horizon path planning.

3.6 Simulation Study

3.6.1 Parameter Selection

For the path planning, it is assumed that the vehicle can sense the environment up to 5 m ahead; we uniformly distribute 6 knots on the 5-meter range and insert 4 samples between knots; choose $o = 4$, $l = 4$ and update the reference path with current 5-meter data whenever the vehicle moves 1 m forward in the x-axis. The tolerances for discontinuity are $\epsilon_1 = \epsilon_2 = \epsilon_3 = 10^{-2}$ (m).

For NMPC parameters, we use $\rho = 1, \delta = 0.1sec$, the prediction horizon $T = 8\delta$. The reference system is chosen according to Remark 3.5. The weighting matrices are selected as $\mathbf{Q} = \text{diag}(\mathrm{q}_{11}, \mathrm{q}_{22}, \mathrm{q}_{33}, \mathrm{q}_{44}, \mathrm{q}_{55}, \mathrm{q}_{66}) = 0.4\,\mathbf{I}_6$, $\mathbf{R} = \text{diag}(\mathrm{r}_{11}, \mathrm{r}_{22}, \mathrm{r}_{33}) = 0.01\mathbf{I}_3$, $\mathrm{K}_1 = \text{diag}(\alpha_1, \beta_1, \gamma_1) = 0.5\mathbf{I}_3$, and $\mathrm{K}_2 = -\text{diag}(\alpha_2, \beta_2, \gamma_2) = -\mathbf{I}_3$. It can be verified that these parameters satisfy (3.18).

3.6.2 Tracking Performance

In the simulations, the optimization Problems 3.1–3.3 are solved by the embedded sequential quadratic programming (SQP) algorithm in the MATLAB function `fmincon`.

The simulation results of the combined path planning and tracking control are shown in Fig. 3.4. It can be observed that the generated reference path is smooth in light of small curvature, which facilitates the tracking control of the AUV. The NMPC controller successfully steers the AUV asymptotically convergent to and then track the reference path in the presence of initial error. The average solution time per NMPC step of is 1.9642 s which needs to be improved for real-time implementation. The computational complexity issue will be further investigated in Chap. 8. Also one can refer to the excellent review paper (Diehl et al. 2009) for possible solutions.

Figure 3.5 shows the generalized control forces and moments, and Fig. 3.6 depicts the transformed real thrust forces. Here, we shall notice that the control inputs are not optimal in terms of fuel consumption (minimum control force) due to the selection of the reference system. As mentioned earlier, unlike the mobile robot whose sway

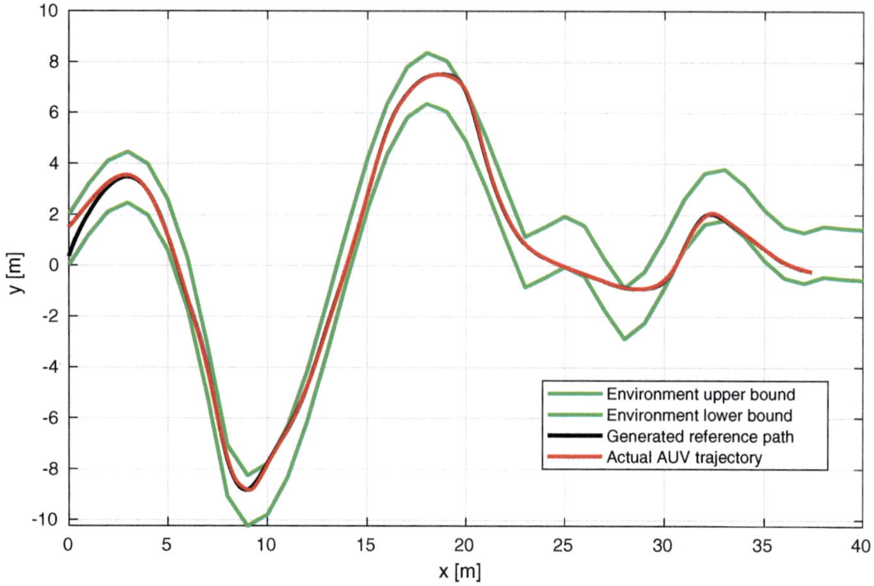

Fig. 3.4 Simulation results of the combined AUV problem

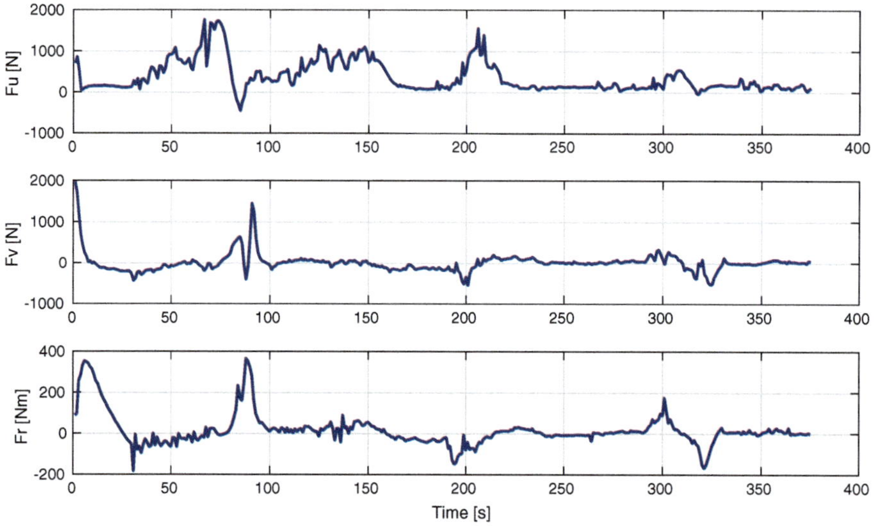

Fig. 3.5 Generalized control signal

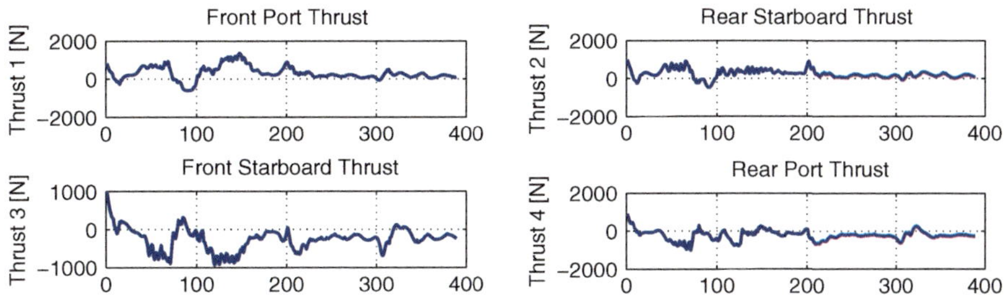

Fig. 3.6 Real thrust forces

velocity v is eliminated by friction, in water, the AUV will always slip sideways when taking turns. Therefore, although we are able to conveniently find the reference system as stated in Remark 3.5, the AUV has to compensate the sway velocity v all the time, which increases the control effort.

Nevertheless, if we consider this point from another perspective, the compensation of v poses a request for the minimum curvature reference path, which is consistent with the proposed path planning method.

3.7 Notes and Summary

In this chapter, we have presented a unified RHO framework for solving the combined path planning and tracking control problem of an AUV. Firstly, a novel RHO-based dynamic path planning method was proposed for the AUV to generate a minimum curvature reference path in the constrained workspace. With the RHO formulation, the dynamic path planning can well accommodate the finite perceiving capability of the vehicle. Then, the NMPC was employed for the AUV to precisely track the planned reference path. Sufficient conditions on closed-loop stability were explored and further applied to guide the NMPC controller design. Finally, an implementation algorithm inherently combining NMPC and the proposed path planning method was developed. With the implementation algorithm, the obtained closed-loop stability of the NMPC tracking control can be ensured. Simulation results on the Falcon dynamic model with the identified model parameters revealed the effectiveness of the proposed control algorithm.

References

Antoniou A, Lu WS (2007) Practical optimization. Springer, New York

Berglund T, Brodnik A, Jonsson H, Staffanson M, Soderkvist I (2010) Planning smooth and obstacle-avoiding B-spline paths for autonomous mining vehicles. IEEE Trans Autom Sci Eng 7(1):167–172

Bingham B, Foley B, Singh H, Camilli R, Delaporta K, Eustice R, Mallios A, Mindell D, Roman C, Sakellariou D (2010) Robotic tools for deep water archaeology: surveying an ancient shipwreck with an autonomous underwater vehicle. J Field Robot 27(6):702–717

Brinón-Arranz L, Seuret A, Canudas-de-Wit C (2014) Cooperative control design for time-varying formations of multi-agent systems. IEEE Trans Autom Control 59(8):2283–2288

Caldwell CV, Dunlap DD, Collins EG (2010) Motion planning for an autonomous underwater vehicle via sampling based model predictive control. In: Proceedings of OCEANS 2010 MTS/IEEE SEATTLE. IEEE, pp 1–6

De Boor C, De Boor C (1978) A practical guide to splines, vol 27. Springer, New York

Diehl M, Ferreau HJ, Haverbeke N (2009) Efficient numerical methods for nonlinear MPC and moving horizon estimation. In: Nonlinear model predictive control. Springer, pp 391–417

Fossen TI (2002) Marine control systems: guidance, navigation, and control of ships, rigs and underwater vehicles. In: Marine cybernetics, Trondheim, Norway

Fossen TI, Pettersen KY (2014) On uniform semiglobal exponential stability (USGES) of proportional line-of-sight guidance laws. Automatica 50(11):2912–2917

Frezza R (1999) Path following for air vehicles in coordinated flight. In: Proceedings of 1999 IEEE/ASME international conference on advanced intelligent mechatronics. IEEE, pp 884–889

Gu D, Hu H (2006) Receding horizon tracking control of wheeled mobile robots. IEEE Trans Control Syst Technol 14(4):743–749

Holzhüter T (1997) LQG approach for the high-precision track control of ships. IEE Proc-Control Theory Appl 144(2):121–127

Jung D, Tsiotras P (2013) On-line path generation for unmanned aerial vehicles using B-spline path templates. J Guidance Control Dyn 36(6):1642–1653

Jung EJ, Yi BJ, Kim W (2015) Kinematic analysis and motion planning for a planar multiarticulated omnidirectional mobile robot. IEEE/ASME Trans Mechatron 20(6):2983–2995

Lindegaard KP (2003) Acceleration feedback in dynamic positioning. Ph.D. thesis

Li W, Xu S, Zhao G, Goh LP (2005) Adaptive knot placement in b-spline curve approximation. Comput-Aided Des 37(8):791–797

Mayne DQ, Rawlings JB, Rao CV, Scokaert PO (2000) Constrained model predictive control: stability and optimality. Automatica 36(6):789–814

Naeem W, Sutton R, Ahmad SM (2004) Pure pursuit guidance and model predictive control of an autonomous underwater vehicle for cable/pipeline tracking. In: Proceedings-institute of marine engineering science and technology part C journal of marine science and environment. Citeseer, pp 25–35

Repoulias F, Papadopoulos E (2007) Planar trajectory planning and tracking control design for underactuated AUVs. Ocean Eng 34(11–12):1650–1667

Saab Seaeye Ltd. (2022) The Falcon ROV. https://www.saabseaeye.com/solutions/underwater-vehicles/falcon

Saccon A, Hauser J, Beghi A (2011) Trajectory exploration of a rigid motorcycle model. IEEE Trans Control Syst Technol 20(2):424–437

Wang L, Zhang LJ, Jia HM, Wang HB (2012) Horizontal tracking control for AUV based on nonlinear sliding mode. In: Proceedings of 2012 IEEE international conference on information and automation. IEEE, pp 460–463

Yu H, Meier K, Argyle M, Beard RW (2014) Cooperative path planning for target tracking in urban environments using unmanned air and ground vehicles. IEEE/ASME Trans Mechatron 20(2):541–552

Chapter 4
Lyapunov-Based Model Predictive Control for Dynamic Positioning and Trajectory-Tracking Control of an AUV

4.1 Introduction

4.1.1 Research Background and Contributions

The DP control is an important motion control for the AUV. Traditionally, the DP functionality refers to the vehicle's capability of reaching and maintaining at a desired position with a desired attitude by exclusively means of active thrusters. In recent studies, it tends to encompass all of the low-speed maneuvering into the definition of DP (Sørensen 2011). The most widely used DP controllers are of the PID type since they are model-free and easy to implement. Acceleration feedback can be articulated with the PID controller to get improved control stability (Fossen et al. 2002). For model-based control, in (Sørensen et al. 1996), the LQG technique is applied to the DP control design with a linearized model. An H_∞ controller is proposed in (Katebi et al. 1997) which makes a trade-off between track keeping and station keeping. Obviously, the DP controllers based on linear control techniques only obtain local stability properties. Nonlinear DP control design, therefore, comes in for better closed-loop properties. The mainstream nonlinear control methods for DP include the Lyapunov-based backstepping and sliding mode control. In (Fossen et al. 1998), the observer-controller structure is investigated for the DP control, and a Lyapunov-based backstepping nonlinear control law is designed. Global exponential stability can be claimed. The nonlinear sliding model DP control can be found in Tannuri et al. (2010). Experiments demonstrate the improved control performance comparing to the conventional PID controller. Most recent DP control designs begin to consider more practical issues. The system constraints such as control limit and safe operating region are inevitable in real-world AUV applications. Therefore, it would be preferred to consider these constraints in the DP controller design.

Y. Shi et al., *Advanced Model Predictive Control for Autonomous Marine Vehicles*, Advances in Industrial Control, https://doi.org/10.1007/978-3-031-19354-5_4

MPC, being an optimization-based time-domain control technique, is known as one of few control methods that can explicitly incorporate constraints into the controller design phase (Mayne et al. 2000; Mayne 2014; Li and Shi 2017). In the context of DP control, the MPC solution is reported in (Veksler et al. 2016) where the simulation results indicate that MPC offers significantly improved control performance. In addition, the TA subproblem, i.e., the coordination of thrusters to generate the matched generalized force and moment requested by the DP control law, can be solved simultaneously with the LMPC-based DP control. Nevertheless, the big question associated with the MPC control design is how to guarantee closed-loop stability, especially for nonlinear systems such as AUVs. The optimality is not necessarily bringing stability to the closed-loop system. To obtain this important closed-loop property, in the standard MPC design procedure, we need to employ additional terminal constraints to the formulated optimization problem and construct a local stabilizing control law via local linearization (Mayne et al. 2000). This essentially introduces considerable conservativeness, and only local stability can be guaranteed. For DP of AUVs, since the system is highly nonlinear, the stability obtained by local linearization may not be practically useful. One potential solution is to circumvent this question by means of LMPC (Christofides et al. 2011; Mhaskar et al. 2006). The LMPC method combines the merits of Lyapunov-based control and model predictive control, making the MPC design relatively simple and intuitive. More importantly, the LMPC controller inherits the stability property of the Lyapunov-based control, which is attractive for the AUV DP control since many existing globally stable DP controllers can be exploited.

In this chapter, we address the DP control problem for an AUV in the local level plane. To handle the highly nonlinear, cross-coupled dynamics and system constraints in terms of thrust limits, the Lyapunov-based nonlinear model predictive control strategy is proposed. Within the LMPC framework, we do not need to solve the TA subproblem separately since the thruster distribution can be formulated into the OCP directly. Considering that most DP controllers are of the PID type, the nonlinear PD control is adopted as the auxiliary control law. The recursive feasibility and closed-loop stability of the LMPC system are explicitly analyzed. Since the nonlinear PD control is globally asymptotically stable, the LMPC-based DP control obtains a quasi-global closed-loop stability: The ROA can be arbitrarily specified by the design parameters.

Furthermore, the obtained LMPC results in DP control are also extended to solve the trajectory-tracking control problem for AUVs. To construct the contraction constraint in the LMPC problem formulation, the Lyapunov-based backstepping procedure is exploited. Sufficient conditions that ensure the recursive feasibility and closed-loop stability are provided analytically. Compared with the NMPC tracking control explored in Chap. 3, the main advantage of the LMPC method is that it builds on the existing AUV control system and utilizes onboard computing resources (online optimization) to improve the control performance. Since the closed-loop stability of LMPC-based control system does not rely on the exact solution of the optimization, we can easily control the computational complexity by specifying the maximum iteration number for the NLP solver. In addition, from the control-theoretic perspective, another advantage of LMPC is that using LMPC we can characterize an explicit ROA, while with the NMPC we can only implicitly define the ROA.

The main contributions of this chapter are summarized as follows:

- A novel LMPC framework is developed for the AUV motion control problems, providing a bridge connecting the modern optimization theory and the conventional marine control technology. The control performance and robustness can be significantly improved by the online optimization.
- Based on the LMPC framework, a DP controller is designed for the AUV. Sufficient conditions for recursive feasibility and closed-loop stability are explicitly derived. With the nonlinear PD control law adopted as the auxiliary controller, we show that the ROA of the LMPC-based DP control system can be specified arbitrarily large.
- An LMPC-based trajectory-tracking controller is also designed. Sufficient conditions for recursive feasibility and closed-loop stability are provided. A guaranteed ROA is characterized analytically.
- With the LMPC framework the TA, subproblem can be solved simultaneously with the AUV motion control, which reduces the conservativeness brought by conventional TA solutions.

4.1.2 Chapter Organization

The remaining part of this chapter is organized as follows: Sect. 4.2 proposes the LMPC framework and details the DP controller design. In Sect. 4.3, the LMPC method is applied to solve the AUV trajectory-tracking problem; controller design and stability analysis are provided. In Sect. 4.4, the conclusion is made for this chapter.

Throughout this chapter, the following notation is used: The column operation $[\rho_1^{\mathrm{T}}, \ldots, \rho_n^{\mathrm{T}}]^{\mathrm{T}}$ is denoted as $\mathrm{col}(\rho_1, \ldots, \rho_n)$; the diagonal operation is denoted by $\mathrm{diag}(\cdot)$; the square of a weighted Euclidean norm $\rho^{\mathrm{T}} A \rho$ is abbreviated by $\|\rho\|_A^2$; and the infinity norm and 2 norm are denoted by $\|\cdot\|_\infty$ and $\|\cdot\|_2$, respectively. The symbol $\mathbf{1}$ is used to represent a column vector with all elements to be 1. The $\max\{\cdot\}$ is a function that returns the largest element within the brace. The absolute value operator applying on a vector $|\rho|$ takes the absolute value for each and every element.

4.2 LMPC Design for Dynamic Positioning Control

4.2.1 Problem Formulation

The AUV model studied for the DP control and the subsequent trajectory-tracking control keeps the same as the one used in Chap. 3 which is established based on the kinematic equations, the dynamic equations, and the thrust distribution:

$$\dot{\mathbf{x}} = \begin{bmatrix} \mathbf{R}(\psi)\mathbf{v} \\ \mathbf{M}^{-1}(\mathbf{Bu} - \mathbf{C}(\mathbf{v})\mathbf{v} - \mathbf{D}(\mathbf{v})\mathbf{v} - \mathbf{g}(\boldsymbol{\eta})) \end{bmatrix} = \mathbf{f}(\mathbf{x}, \mathbf{u}), \tag{4.1}$$

where the state vector $\mathbf{x} = [x, y, \psi, u, v, r]^{\mathrm{T}}$ is consisted of the pose and velocity of the vehicle, and the control vector $\mathbf{u} = [u_1, u_2, u_3, u_4]^{\mathrm{T}}$ is consisted of the forces generated by the four thrusters. The detailed expression can be found in (2.32) and (2.34).

The DP control refers to the feedback control of marine vehicles to reach and maintain the desired position at the preferred orientation by exclusively means of thrusters. Consider the desired position and orientation given by $\boldsymbol{\eta}_d = [x_d, y_d, \psi_d]^{\mathrm{T}}$. An intuitive MPC formulation can be established for the DP control

$$\begin{aligned} \min_{\hat{\mathbf{u}} \in S(\delta)} \ & J = \int_0^T (\|\tilde{\mathbf{x}}(s)\|_Q^2 + \|\hat{\mathbf{u}}(s)\|_R^2) ds + \|\tilde{\mathbf{x}}(T)\|_P^2 \\ \text{s.t.} \ & \dot{\hat{\mathbf{x}}}(s) = \mathbf{f}(\hat{\mathbf{x}}(s), \hat{\mathbf{u}}(s)), \\ & \hat{\mathbf{x}}(0) = \mathbf{x}(t_0), \\ & |\hat{\mathbf{u}}(s)| \leq \mathbf{u}_{\max}, \end{aligned} \tag{4.2}$$

where $\hat{\mathbf{x}}(s)$ is the predicted state trajectory of the vehicle, evolving using the system model; $\tilde{\mathbf{x}} = \mathrm{col}(\tilde{\boldsymbol{\eta}}, \hat{\mathbf{v}})$ denotes the error state in which $\tilde{\boldsymbol{\eta}} = \hat{\boldsymbol{\eta}} - \boldsymbol{\eta}_d$; $S(\delta)$ denotes the family of piecewise constant functions characterized by the sampling period δ. $T = N\delta$ is the prediction horizon, and the weighting matrices Q, R, P are positive definite.

However, even if the global optimal solution curve $\hat{\mathbf{u}}^*(s)$ can be obtained, the closed-loop stability is not guaranteed due to the finite prediction horizon. A complex offline design procedure appears to be a necessity. For nonlinear systems, basically, local linearization with respect to the equilibrium point has to take place to facilitate the choice of weighting matrices as well as the construction of the auxiliary local feedback control (Mayne 2014). Therefore, only local stability property can be claimed.

In view of the fact that many existing DP controllers are designed using the Lyapunov direct method and obtain the global stability property, we can explicitly take advantage of these existing controllers to formulate the LMPC problem (P_0) for the DP control

$$\min_{\hat{\mathbf{u}} \in S(\delta)} J = \int_0^T (\|\tilde{\mathbf{x}}(s)\|_Q^2 + \|\hat{\mathbf{u}}(s)\|_R^2) ds + \|\tilde{\mathbf{x}}(T)\|_P^2 \tag{4.3a}$$

$$\text{s.t. } \dot{\hat{\mathbf{x}}}(s) = \mathbf{f}(\hat{\mathbf{x}}(s), \hat{\mathbf{u}}(s)), \tag{4.3b}$$

$$\hat{\mathbf{x}}(0) = \mathbf{x}(t_0), \tag{4.3c}$$

$$|\hat{\mathbf{u}}(s)| \leq \mathbf{u}_{\max}, \tag{4.3d}$$

$$\frac{\partial V}{\partial \mathbf{x}}\mathbf{f}(\hat{\mathbf{x}}(0), \hat{\mathbf{u}}(0)) \leq \frac{\partial V}{\partial \mathbf{x}}\mathbf{f}(\hat{\mathbf{x}}(0), h(\hat{\mathbf{x}}(0))), \tag{4.3e}$$

where $h(\cdot)$ is the existing DP controller and $V(\cdot)$ is the corresponding Lyapunov function. The presence of contraction constraint (4.3e) allows us to show that the LMPC inherits the stability properties of the state feedback control $h(\mathbf{x})$ (Christofides et al. 2011). Therefore, if $h(\mathbf{x})$ brings the global stability of desired equilibrium point, the system under LMPC control is also globally stable. Meanwhile, thanks to the online optimization procedure, the LMPC controller can improve the DP control performance considerably.

The LMPC-based DP control algorithm will be implemented in the standard receding horizon control fashion: (i) At the current sampling instant t_0, given the system state $\mathbf{x}(t_0)$, the OCP (P_0) is solved; let $\boldsymbol{\kappa}(s)$ denote the (sub-)optimal solution; (ii) the vehicle implements $\boldsymbol{\kappa}(s)$ for only one sampling period: $\mathbf{u}(t) = \boldsymbol{\kappa}(s)$ for $s \in [0, \delta]$; (iii) at next sampling instant $t_0 + \delta$, new measurement of system state $\mathbf{x}(t_0 + \delta)$ is fed back, and (P_0) will be solved again with the new initial condition $\mathbf{x}(t_0 + \delta)$. Then repeat from (i).

4.2.2 Main Results

Since the universally used DP controllers are of the PID type, we investigate the multivariable PD control for the purpose of constructing the contraction constraint in the LMPC problem (4.3). However, in principle, any other Lyapunov-based nonlinear controller can be used.

Consider the following nonlinear PD control law:

$$\begin{aligned} \boldsymbol{\tau}(\mathbf{x}) &= \mathbf{g}(\boldsymbol{\eta}) - \mathbf{R}^{\mathrm{T}}(\psi)\boldsymbol{\tau}_{PD}, \\ \boldsymbol{\tau}_{PD} &= \mathbf{K}_p\tilde{\boldsymbol{\eta}} + \mathbf{K}_d\dot{\boldsymbol{\eta}}, \end{aligned} \tag{4.4}$$

where $\mathbf{K}_p$ and $\mathbf{K}_d$ are the user-specified control gain matrices that are diagonal and positive definite.

The Lyapunov function candidate is suggested as follows:

$$V = \frac{1}{2}\mathbf{v}^{\mathrm{T}}\mathbf{M}\mathbf{v} + \frac{1}{2}\tilde{\boldsymbol{\eta}}^{\mathrm{T}}\mathbf{K}_p\tilde{\boldsymbol{\eta}}. \tag{4.5}$$

Taking time derivative of V along the trajectory of the closed-loop system, we have

$$\dot{V} = \mathbf{v}^{\mathrm{T}}\mathbf{M}\dot{\mathbf{v}} + \dot{\boldsymbol{\eta}}^{\mathrm{T}}\mathbf{K}_p\tilde{\boldsymbol{\eta}}. \tag{4.6}$$

Substituting (2.31), (2.33), and (4.4) into (4.6) yields

$$\dot{V} = -\mathbf{v}^{\mathrm{T}}[\mathbf{C}(\mathbf{v}) + \mathbf{D}(\mathbf{v}) + \mathbf{K}_d^*(\boldsymbol{\eta})]\mathbf{v}, \tag{4.7}$$

where $\mathbf{K}_d^*(\boldsymbol{\eta}) = \mathbf{R}^{\mathrm{T}}(\psi)\mathbf{K}_d\mathbf{R}(\psi)$. Considering $\mathbf{v}^{\mathrm{T}}\mathbf{C}(\mathbf{v})\mathbf{v} = 0$ for all $\mathbf{v}$, we have

$$\dot{V} = -\mathbf{v}^{\mathrm{T}}[\mathbf{D}(\mathbf{v}) + \mathbf{K}_d^*(\boldsymbol{\eta})]\mathbf{v} \le 0, \tag{4.8}$$

since it can be easily shown that $\mathbf{K}_d^*(\boldsymbol{\eta}) > 0$. By LaSalle's theorem (Khalil 1996), the closed-loop system with the nonlinear PD controller is globally asymptotically stable with respect to the equilibrium $[\tilde{\boldsymbol{\eta}}, \mathbf{v}] = [\mathbf{0}, \mathbf{0}]$.

Then the detailed expression of the contraction constraint (4.3e) corresponding to the nonlinear PD control is

$$\begin{aligned} &\hat{\mathbf{v}}(0)^{\mathrm{T}}(\hat{\mathbf{u}}(0) - \mathbf{C}(\hat{\mathbf{v}}(0))\hat{\mathbf{v}}(0) - \mathbf{D}(\hat{\mathbf{v}}(0))\hat{\mathbf{v}}(0) - \mathbf{g}(\hat{\boldsymbol{\eta}}(0)) + \mathbf{R}^{\mathrm{T}}(\hat{\psi}(0))\mathbf{K}_p\tilde{\boldsymbol{\eta}}(0)) \\ &\le -\hat{\mathbf{v}}(0)^{\mathrm{T}}[\mathbf{D}(\hat{\mathbf{v}}(0)) + \mathbf{K}_d^*(\hat{\boldsymbol{\eta}}(0))]\hat{\mathbf{v}}(0). \end{aligned} \tag{4.9}$$

For the recursive feasibility, we notice that the PD controller $h(\hat{\mathbf{x}})$ is always feasible for the LMPC problem (4.3) provided that $|h(\hat{\mathbf{x}})| \le \mathbf{u}_{\max}$ can be satisfied.

For calculation simplicity, in the following we make several reasonable and practical assumptions.

Assumption 4.1 The thrusters have the same maximum capacity, i.e., $|u_i| \le u_{\max}$.

Note that Assumption 4.1 is reasonable and often true in practice. Then we have the following proposition:

Proposition 4.1 *Consider the TA using the Moore–Penrose pseudoinverse implementation, i.e.,*

$$\mathbf{u} = \mathbf{B}^{\mathrm{T}}(\mathbf{B}\mathbf{B}^{\mathrm{T}})^{-1}\boldsymbol{\tau} = \mathbf{B}^{+}\boldsymbol{\tau}, \tag{4.10}$$

and denote the maximum possible generalized thrust force by $\tau_{\max} = \|\boldsymbol{\tau}_{\max}\|_\infty$ *with* $\boldsymbol{\tau}_{max} = [F_{u,\max}, F_{v,\max}, F_{r,\max}]^{\mathrm{T}}$.

If the following relation holds:

$$\tau_{\max} \le \frac{u_{\max}}{\bar{b}^{+}}, \tag{4.11}$$

where $\bar{b}^{+} = \|\mathbf{B}^{+}\|_\infty$*, then the TA is always feasible, i.e.,* $\|\mathbf{u}\|_\infty \le u_{\max}$.

Proof Taking the infinity norm on both sides of (4.10) we have

$$\|\mathbf{u}\|_\infty = \|\mathbf{B}^{+}\boldsymbol{\tau}\|_\infty \le \bar{b}^{+}\|\boldsymbol{\tau}\|_\infty \le \bar{b}^{+}\tau_{\max}. \tag{4.12}$$

Having (4.11) and Assumption 4.1, it follows

$$\|\mathbf{u}\|_\infty \le \bar{b}^{+}\tau_{\max} \le u_{\max}. \tag{4.13}$$

□

Assumption 4.2 The restoring force $\mathbf{g}(\boldsymbol{\eta})$ is bounded and relatively small such that

$$\|\mathbf{g}(\boldsymbol{\eta})\|_\infty \le \bar{g} < \tau_{\max}, \tag{4.14}$$

where $\bar{g}$ denotes the bound.

The Assumption 4.2 is also reasonable. The detailed expression of $\mathbf{g}(\boldsymbol{\eta})$ can be found in (2.24) which contains combinations of sine and cosine functions. Therefore, the boundedness of the restoring force can be guaranteed. Furthermore, the bound $\bar{g}$ is relatively small with respect to the maximum allowed thrust force $\tau_{\max}$. If not, from (4.4) we see that there would be no room for the feedback control, which is not considered in this book. Then we have the following theorem:

Theorem 4.1 *Suppose the control gains* $\mathbf{K}_p = \text{diag}(k_{pi})$ *and* $\mathbf{K}_d = \text{diag}(k_{di})$. *Let* $\bar{k}_p = \max\{k_{pi}\}$ *denote the largest element in* $\mathbf{K}_p$ *and* $\bar{k}_d = \max\{k_{di}\}$ *denote the largest element in* $\mathbf{K}_d$. *Suppose Assumptions 4.1 and 4.2 hold, and define* $h(\mathbf{x}) = \mathbf{B}^{+}\boldsymbol{\tau}(\mathbf{x})$. *If the following relation holds:*

$$(\bar{k}_p + \sqrt{2}\bar{k}_d)\|\tilde{\mathbf{x}}(0)\|_2 \leq \frac{\tau_{\max} - \bar{g}}{\sqrt{2}}, \tag{4.15}$$

where $\tilde{\mathbf{x}}(0)$ *is the initial error and* $\tau_{\max}$ *follows (4.11), then the LMPC* (P_0) *admits recursive feasibility, i.e.,* $|h(\hat{\mathbf{x}}(t))| \leq \mathbf{u}_{\max}$ *for all* $t \geq 0$ *where* $\mathbf{u}_{\max} = u_{\max}\mathbf{1}$.

Proof Taking infinity norm on both sides of (4.4) yields

$$\begin{aligned} \|\boldsymbol{\tau}\|_\infty &= \|\mathbf{g}(\boldsymbol{\eta}) - \mathbf{R}^{\mathrm{T}}(\psi)\boldsymbol{\tau}_{PD}\|_\infty \\ &\leq \|\mathbf{g}(\boldsymbol{\eta})\|_\infty + \|\mathbf{R}^{\mathrm{T}}(\psi)\|_\infty\|\boldsymbol{\tau}_{PD}\|_\infty \\ &\leq \bar{g} + \sqrt{2}\|\boldsymbol{\tau}_{PD}\|_\infty, \end{aligned} \tag{4.16}$$

since $\|\mathbf{R}^{\mathrm{T}}(\psi)\|_\infty = \max\{\cos\psi - \sin\psi, \sin\psi + \cos\psi, 1\} \leq \sqrt{2}$.

From (2.33) and (4.4), we have

$$\begin{aligned} \|\boldsymbol{\tau}_{PD}\|_\infty &= \|\mathbf{K}_p\tilde{\boldsymbol{\eta}} + \mathbf{K}_d\dot{\boldsymbol{\eta}}\|_\infty = \|\mathbf{K}_p\tilde{\boldsymbol{\eta}} + \mathbf{K}_d\mathbf{R}(\psi)\mathbf{v}\|_\infty \\ &\leq \bar{k}_p\|\tilde{\boldsymbol{\eta}}\|_\infty + \sqrt{2}\bar{k}_d\|\mathbf{v}\|_\infty \\ &\leq (\bar{k}_p + \sqrt{2}\bar{k}_d)\|\tilde{\mathbf{x}}\|_\infty. \end{aligned} \tag{4.17}$$

Since (4.3e) is satisfied, it admits $\dot{V} \leq 0$. Therefore, $\|\tilde{\mathbf{x}}\|_2 \leq \|\tilde{\mathbf{x}}(0)\|_2$. Considering $\|\tilde{\mathbf{x}}\|_\infty \leq \|\tilde{\mathbf{x}}\|_2$, we have

$$\|\boldsymbol{\tau}_{PD}\|_\infty \leq (\bar{k}_p + \sqrt{2}\bar{k}_d)\|\tilde{\mathbf{x}}(0)\|_2. \tag{4.18}$$

Together with (4.16), we have

$$\|\boldsymbol{\tau}\|_\infty \leq \bar{g} + \sqrt{2}(\bar{k}_p + \sqrt{2}\bar{k}_d)\|\tilde{\mathbf{x}}(0)\|_2. \tag{4.19}$$

If (4.15) can be satisfied, then the following relation holds:

$$\|\boldsymbol{\tau}\|_\infty \leq \bar{g} + \sqrt{2}(\bar{k}_p + \sqrt{2}\bar{k}_d)\|\tilde{\mathbf{x}}(0)\|_2 \leq \tau_{\max}. \tag{4.20}$$

With (4.11), we can guarantee that $\|h(\hat{\mathbf{x}}(t))\|_\infty \leq u_{\max}$ is satisfied all the time, which completes the proof. □

We notice that the (4.15) can be easily satisfied since $\bar{k}_p$ and $\bar{k}_d$ can be specified as arbitrarily small positive numbers. As the recursive feasibility implies the closed-loop stability, the ROA can be arbitrarily large.

Definition 4.1 (Class $\mathcal{K}_\infty$ Function) A continuous function $\alpha : [0, \infty) \to [0, \infty)$ is said to belong to $\mathcal{K}_\infty$ if

- it is strictly increasing;
- it is such that $\alpha(0) = 0$ and $\lim_{r\to\infty} \alpha(r) = \infty$.

Theorem 4.2 *Suppose Assumptions 4.1 and 4.2 hold, then the LMPC-based DP control makes the desired equilibrium point* $[\tilde{\boldsymbol{\eta}}, \mathbf{v}] = [\mathbf{0}, \mathbf{0}]$ *asymptotically stable. Furthermore, the ROA can be arbitrarily large with enough small control gains* $\bar{k}_p$ *and* $\bar{k}_d$.

Proof The proof first shows that the equilibrium is asymptotically stable and then illustrates that the ROAcan be arbitrarily large.

Since we have already found a Lyapunov function $V(\mathbf{x})$ in (4.5), continuously differentiable and radically unbounded, by converse Lyapunov theorems (Khalil 1996), there exist functions $\alpha_i(\cdot), i = 1, 2, 3$ belonging to class $\mathcal{K}_\infty$ such that the following inequalities hold:

$$\alpha_1(\|\mathbf{x}\|) \leq V(\mathbf{x}) \leq \alpha_2(\|\mathbf{x}\|), \tag{4.21a}$$

$$\frac{\partial V}{\partial \mathbf{x}}\mathbf{f}(\mathbf{x}, h(\mathbf{x})) \leq -\alpha_3(\|\mathbf{x}\|). \tag{4.21b}$$

Considering (4.3e) and that only the first element of $\boldsymbol{\kappa}(\mathbf{x})$ will be implemented for each sampling period, we have

$$\frac{\partial V}{\partial \mathbf{x}}\mathbf{f}(\mathbf{x}, \mathbf{u}(\mathbf{x})) \leq \frac{\partial V}{\partial \mathbf{x}}\mathbf{f}(\mathbf{x}, h(\mathbf{x})) \leq -\alpha_3(\|\mathbf{x}\|). \tag{4.22}$$

By standard Lyapunov arguments [e.g., Khalil (1996, Theorem 4.8)] we claim that the closed-loop system with the LMPC control $\mathbf{u}(\mathbf{x})$ is asymptotically stable with a region of attraction

$$\mathcal{X} = \{\mathbf{x} \in \mathbb{R}^n \mid (\bar{k}_p + \sqrt{2}\bar{k}_d)\|\tilde{\mathbf{x}}\|_2 \leq \frac{\tau_{\max} - \bar{g}}{\sqrt{2}}\}, \tag{4.23}$$

where $\tilde{\mathbf{x}} = \text{col}(\tilde{\boldsymbol{\eta}}, \mathbf{v})$ denotes the error state.

Obviously, with arbitrarily large initial error $\tilde{\mathbf{x}}$, we choose the control gains $\mathbf{K}_p > 0$ and $\mathbf{K}_d > 0$ satisfying

$$(\bar{k}_p + \sqrt{2}\bar{k}_d) \leq \frac{\tau_{\max} - \bar{g}}{\sqrt{2}\|\tilde{\mathbf{x}}\|_2}. \tag{4.24}$$

Then the LMPC problem is feasible and the closed-loop system is stable. Since there are no other constraints on $\bar{k}_p$ and $\bar{k}_d$, the ROA can be arbitrarily large with enough small control gains satisfying (4.24). □

Remark 4.1 Although the asymptotic stability relies only on the positive definiteness of the control gain matrices $\mathbf{K}_p$ and $\mathbf{K}_d$, the control performance of the PD controller is determined by the magnitude of the control gains. Smaller control gains will result in slower convergence. However, for the proposed LMPC-based DP control, due to the optimization procedure, it can automatically make full use of the thrust capability to generate the best achievable control performance, with respect to the objective function (4.3a), even though we have selected very small control gains for a large ROA.

4.2.3 Simulation Study

In this section, the proposed LMPC-based DP control is simulated on the dynamic model of Falcon, which is identified through experimental data (Proctor 2014). The AUV model parameters are the same as those in the previous chapter, and the details have been summarized in Table 2.1.

Without loss of generality, the desired position is chosen to be the origin of the IRF, i.e., $\boldsymbol{\eta}_d = [0, 0, 0]^{\mathrm{T}}$. The LMPC problem (4.3) is discretized and then solved by the sequential SQP method.

The controller parameters are chosen as follows: The sampling period $\delta = 0.1$ s, prediction horizon $T = 5\delta$, and the weighting matrices $Q = \mathrm{diag}(10^5, 10^5, 10^4, 10^3, 10^3, 10^3)$, $R = \mathrm{diag}(10^{-3}, 10^{-3}, 10^{-3}, 10^{-3})$, and $P = \mathrm{diag}(10^3, 10^3, 10^2, 10, 10, 10)$. The nonlinear PD control gains $\mathbf{K}_p = \mathbf{K}_d = \mathrm{diag}(10, 10, 10)$. The initial condition is $\mathbf{x}(0) = [5, 5, -\pi/2, 0, 0, 0]^{\mathrm{T}}$.

The simulated AUV trajectories are shown in Fig. 4.1, and the state trajectories with respect to time are illustrated in Fig. 4.2.

As we can see, both LMPC and PD controller steer the AUV successfully to the desired set-point. However, the LMPC-based DP control converges much faster (in about 20 sec.) than the PD control (in about 40 sec.). This is because we have selected relatively small control gain matrices $\mathbf{K}_p$ and $\mathbf{K}_d$ for a large region of attraction. The simulation results demonstrate the superior DP control performance; the resulting improved performance is mainly due to the online optimization. The thrust forces generated by each thruster are plotted in Fig. 4.3. We shall notice that at the beginning of the LMPC control, the thrusters are operated at full speed in order to get the fastest possible convergence rate.

The receding horizon mechanism introduces feedback into the closed-loop system. Therefore, an additional merit of LMPC-based DP control system is the inherent robustness to model uncertainties and external disturbances (Pannocchia et al.

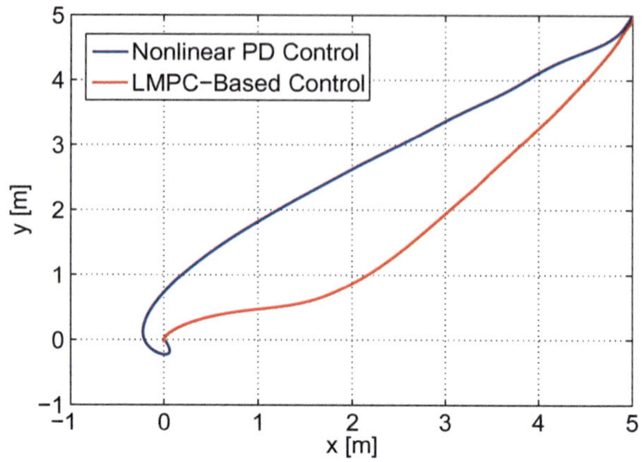

Fig. 4.1 AUV trajectory in local level plane

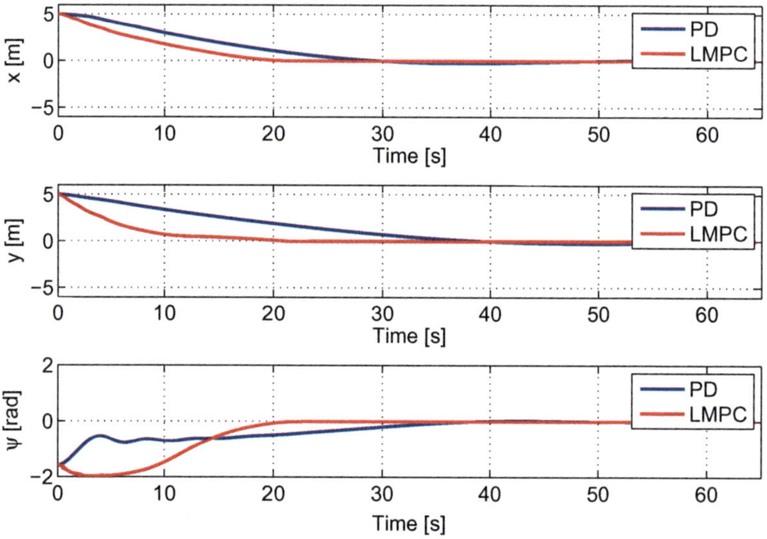

Fig. 4.2 State trajectories

2011), which is attractive especially for marine control systems. The robustness of the LMPC-based DP control is investigated through simulations. In the following simulations, we assume that the system model is subject to 20% model error, and there is a disturbance $\mathbf{w} = [10\text{N}, 10\text{N}, 0\text{Nm}]^{\text{T}}$ exerting on the vehicle all the time, caused by an irrotational ocean current.

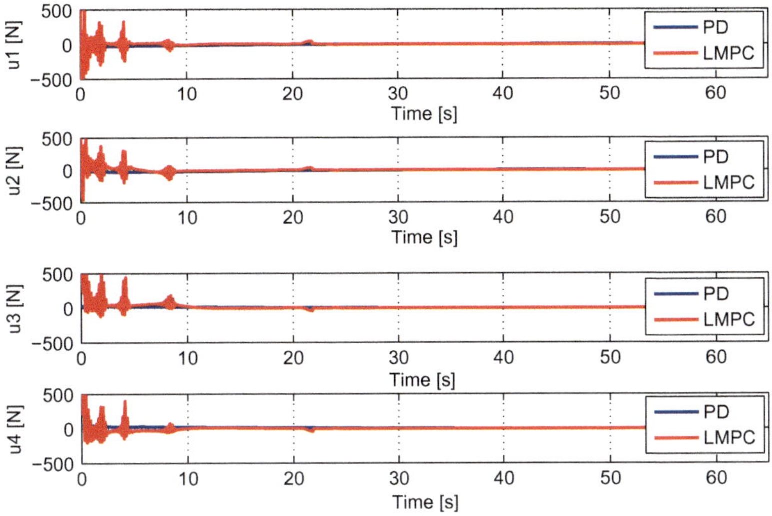

Fig. 4.3 Control input signals

Fig. 4.4 AUV trajectory in local level plane (with disturbance)

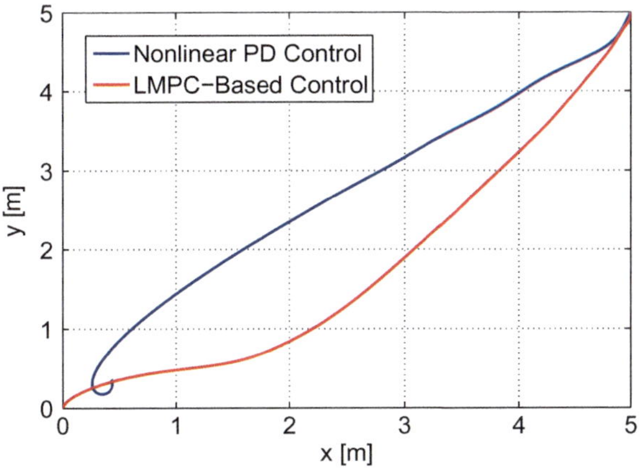

From the simulation results illustrated in Figs. 4.4, 4.5, and 4.6, we see that the LMPC-based DP control still gets converged to the desired position. In contrast, it can be proved that the PD control only ensures uniform ultimate boundedness (UUB) of the position error (Fossen et al. 2002). This has demonstrated that the robustness of DP control can be improved by the introduced online optimization.

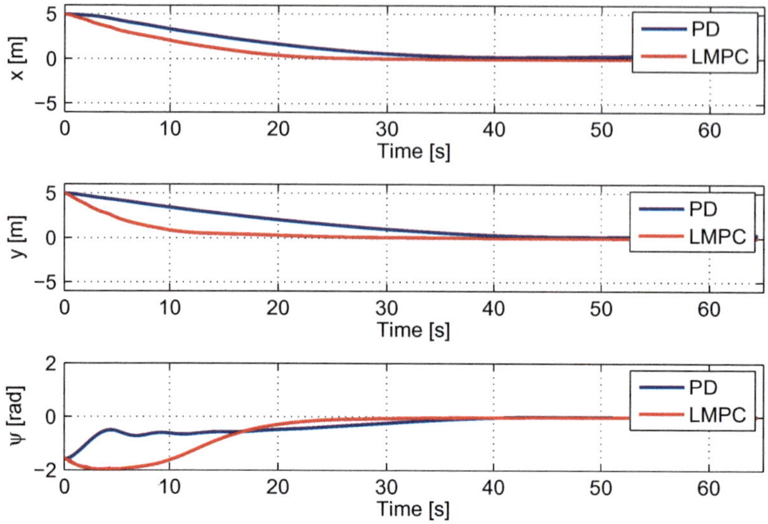

Fig. 4.5 State trajectories (with disturbance)

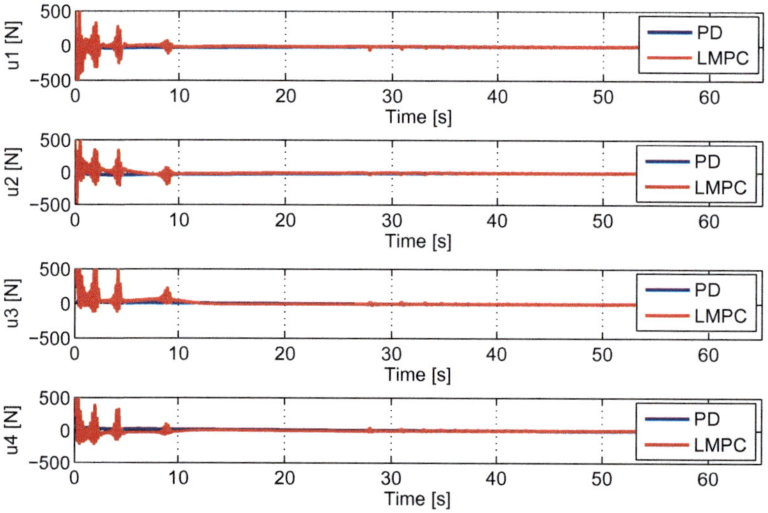

Fig. 4.6 Control input signals (with disturbance)

4.3 LMPC Design for Trajectory-Tracking Control

In this section, we extend the main results obtained in Sect. 4.2 to solve the trajectory-tracking control problem of an AUV.

4.3.1 Problem Formulation

Consider a desired trajectory $p(t) = [x_d(t), y_d(t)]^{\mathrm{T}}$ that defines the position of the AUV in the local level plane. Here we assume that the following conditions hold:

Assumption 4.3 The desired trajectory $p(t)$ and its derivatives are smooth and bounded, satisfying that: $0 \leq \underline{p} \leq \|p(t)\|_\infty \leq \bar{p} < \infty$, $0 < \underline{p_1} \leq \|\dot{p}(t)\|_\infty \leq \bar{p}_1 < \infty$, $0 \leq \underline{p_2} \leq \|\ddot{p}(t)\|_\infty \leq \bar{p}_2 < \infty$ and $0 \leq \underline{p_3} \leq \|\dddot{p}(t)\|_\infty \leq \bar{p}_3 < \infty$.

We augment $p(t)$ to a reference system so that each state of the AUV system (4.1) has a feasible reference. Let $\mathbf{x}_d(t) = [x_d(t), y_d(t), \psi_d(t), u_d(t), v_d(t), r_d(t)]^{\mathrm{T}}$ with

$$
\begin{aligned}
\psi_d(t) &= \operatorname{atan2}(\dot{y}_d(t), \dot{x}_d(t)),\\
u_d(t) &= \sqrt{\dot{x}_d^2(t) + \dot{y}_d^2(t)},\\
v_d(t) &= 0,\\
r_d(t) &= (\dot{x}_d(t)\ddot{y}_d(t) - \dot{y}_d(t)\ddot{x}_d(t))/(\dot{x}_d^2(t) + \dot{y}_d^2(t)),
\end{aligned}
\tag{4.25}
$$

where atan2 is the four-quadrant inverse tangent operator. Then it can be verified that $\mathbf{x}_d(t)$ satisfies the kinematic equations (2.33). Similarly, we assume $\mathbf{v}_d = [u_d, v_d, r_d]^{\mathrm{T}}$ obeys the dynamic equations (2.31), and the reference control forces $\boldsymbol{\tau}_d = [F_{ud}, F_{vd}, F_{rd}]^{\mathrm{T}}$ can be obtained by

$$
\boldsymbol{\tau}_d = \mathbf{M}\dot{\mathbf{v}}_d + \mathbf{C}(\mathbf{v}_d)\mathbf{v}_d + \mathbf{D}(\mathbf{v}_d)\mathbf{v}_d + \mathbf{g}(\boldsymbol{\eta}_d), \tag{4.26}
$$

where $\boldsymbol{\eta}_d = [x_d, y_d, \psi_d]^{\mathrm{T}}$ and $\dot{\mathbf{v}}_d$ can be calculated by taking time derivative of (4.25). Furthermore, since the model property P-5 is satisfied we take advantage of the Moore–Penrose pseudoinverse implementation to solve the TA subproblem and we can get the reference control for each thruster:

$$
\mathbf{u}_d = (\mathbf{B}^{\mathrm{T}}\mathbf{B})^{-1}\mathbf{B}^{\mathrm{T}}\boldsymbol{\tau}_d = \mathbf{B}^{+}\boldsymbol{\tau}_d. \tag{4.27}
$$

Proposition 4.2 *Provided that Assumption 4.3 holds and the reference signals are chosen as (4.25) then the desired pose $\boldsymbol{\eta}_d$ and its derivatives $\dot{\boldsymbol{\eta}}_d$ and $\ddot{\boldsymbol{\eta}}_d$ are upper bounded, i.e., $\|\boldsymbol{\eta}_d(t)\|_\infty \leq \bar{\eta}_d$, $\|\dot{\boldsymbol{\eta}}_d(t)\|_\infty \leq \bar{\eta}_{d1}$ and $\|\ddot{\boldsymbol{\eta}}_d(t)\|_\infty \leq \bar{\eta}_{d2}$ for some positive numbers $\bar{\eta}_d$, $\bar{\eta}_{d1}$ and $\bar{\eta}_{d2}$.*

Proof With Assumption 4.3, we notice that the bounds of $\boldsymbol{\eta}_d$, $\dot{\boldsymbol{\eta}}_d$, and $\ddot{\boldsymbol{\eta}}_d$ depend on the bounds of ψ_d, $\dot{\psi}_d$, and $\ddot{\psi}_d$, respectively.

By definition, we have $|\psi_d| \leq \pi$. Since $\dot{\psi}_d = r_d$, we have $|\dot{\psi}_d| \leq \bar{p}_1\bar{p}_2/\underline{p_1}^2$. We explicit $\ddot{\psi}_d$ as follows:

$$\ddot{\psi}_d = \frac{\dot{x}_d\dddot{y}_d - \dot{y}_d\dddot{x}_d}{\dot{x}_d^2 + \dot{y}_d^2} - \frac{2(\dot{x}_d\ddot{x}_d + \dot{y}_d\ddot{y}_d)(\dot{x}_d\ddot{y}_d - \dot{y}_d\ddot{x}_d)}{(\dot{x}_d^2 + \dot{y}_d^2)^2}. \tag{4.28}$$

Obviously, we have $|\ddot{\psi}_d| \leq \bar{p}_1\bar{p}_3/\underline{p_1}^2 + 2\bar{p}_1^2\bar{p}_2^2/\underline{p_1}^2$. Therefore, the upper bounds can be calculated by

$$\bar{\eta}_d = \max\{\bar{p}, \pi\}, \tag{4.29a}$$

$$\bar{\eta}_{d1} = \max\{\bar{p}_1,\ \bar{p}_1\bar{p}_2/\underline{p_1}^2\}, \tag{4.29b}$$

$$\bar{\eta}_{d2} = \max\{\bar{p}_2,\ \bar{p}_1\bar{p}_3/\underline{p_1}^2 + 2\bar{p}_1^2\bar{p}_2^2/\underline{p_1}^2\}. \tag{4.29c}$$

□

Since $\dot{\boldsymbol{\eta}}_d = \mathbf{R}(\psi)\mathbf{v}_d$, together with (4.26), (4.27) and the model property P-6, it can be easily shown that the reference state $\mathbf{x}_d$ and control $\mathbf{u}_d$ are also finite and upper bounded.

Then the intuitive MPC formulation for the AUV trajectory-tracking control can be safely established

$$\begin{aligned} \min_{\hat{\mathbf{u}} \in S(\delta)} \ & J = \int_0^T (\|\tilde{\mathbf{x}}(s)\|_Q^2 + \|\tilde{\mathbf{u}}(s)\|_R^2) ds + \|\tilde{\mathbf{x}}(T)\|_P^2 \\ \text{s.t.} \ & \dot{\hat{\mathbf{x}}}(s) = \mathbf{f}(\hat{\mathbf{x}}(s), \hat{\mathbf{u}}(s)), \\ & \hat{\mathbf{x}}(0) = \mathbf{x}(t_0), \\ & |\hat{\mathbf{u}}(s)| \leq \mathbf{u}_{\max}, \end{aligned} \tag{4.30}$$

where $\hat{\mathbf{x}}(s)$ is the predicted state trajectory of the vehicle with respect to the predictive control $\hat{\mathbf{u}}(s)$, evolving from $\mathbf{x}(t_0)$ using the system model; $\tilde{\mathbf{x}} = \hat{\mathbf{x}} - \mathbf{x}_d$ is the error state and $\tilde{\mathbf{u}} = \hat{\mathbf{u}} - \mathbf{u}_d$ is the control error; $S(\delta)$ denotes the family of piecewise constant functions characterized by the sampling period δ and $T = N\delta$ is the prediction horizon; The weighting matrices Q, R, and P are positive definite.

As usual, the closed-loop stability cannot be automatically guaranteed by the optimality of the solution due to the finite prediction horizon. To ensure this important closed-loop property, a complex offline design procedure would take place. For nonlinear systems such as the AUV, with standard MPC design technique, local linearization needs to be performed to select appropriate weighting matrices and construct an auxiliary local feedback control law. In this way, the ROA will be defined implicitly (Rawlings et al. 2017).

For the AUV tracking of a curved reference trajectory, the local linearization appears inappropriate to apply. To circumvent the local linearization while ensuring the closed-loop stability of the MPC tracking control, we exploit an auxiliary

Lyapunov-based nonlinear tracking control law and formulate the LMPC problem by adding a contraction constraint to the original MPC formulation:

$$\min_{\hat{\mathbf{u}} \in S(\delta)} J = \int_0^T (\|\tilde{\mathbf{x}}(s)\|_Q^2 + \|\tilde{\mathbf{u}}(s)\|_R^2) ds + \|\tilde{\mathbf{x}}(T)\|_P^2 \tag{4.31a}$$

$$\text{s.t. } \dot{\hat{\mathbf{x}}}(s) = \mathbf{f}(\hat{\mathbf{x}}(s), \hat{\mathbf{u}}(s)), \tag{4.31b}$$

$$\hat{\mathbf{x}}(0) = \mathbf{x}(t_0), \tag{4.31c}$$

$$|\hat{\mathbf{u}}(s)| \leq \mathbf{u}_{\max}, \tag{4.31d}$$

$$\tfrac{\partial V}{\partial \mathbf{x}} \mathbf{f}(\hat{\mathbf{x}}(0), \hat{\mathbf{u}}(0)) \leq \tfrac{\partial V}{\partial \mathbf{x}} \mathbf{f}(\hat{\mathbf{x}}(0), h(\hat{\mathbf{x}}(0))), \tag{4.31e}$$

where $h(\cdot)$ is the auxiliary Lyapunov-based nonlinear tracking control law and $V(\cdot)$ is the corresponding Lyapunov function. The presence of contraction constraint (4.31e) allows us to show that the LMPC controller inherits the stability properties of the state feedback control $h(\mathbf{x})$ and a guaranteed ROA can be explicitly characterized. Furthermore, thanks to the online optimization procedure, the LMPC controller will automatically perform the best achievable tracking control and respect the physical limitation of the system.

The LMPC-based trajectory-tracking control will be implemented in the standard receding horizon fashion and the control algorithm is summarized in Algorithm 4.1.

Algorithm 4.1 LMPC Algorithm

1: Input the objective function J in (4.31a).
2: Measure the current state $\mathbf{x}(t)$.
3: Solve the LMPC problem (4.31) with $\mathbf{x}(t_0) = \mathbf{x}(t)$; let $\boldsymbol{\kappa}(s)$ denote the optimal solution.
4: Implement $\boldsymbol{\kappa}(s)$ for only one sampling period: $\mathbf{u}(t) = \boldsymbol{\kappa}(s)$ for $s \in [0, \delta]$;
5: At next sampling time instant, set $t = t + \delta$, then repeat from step 2.

Remark 4.2 As will be seen shortly in the next section, neither the recursive feasibility nor the closed-loop stability relies on the exact solution of the optimization. In Algorithm 4.1 suboptimal solutions are practically acceptable, which is highly desirable for any nonlinear MPC algorithm by considering the following facts: First, since the system model (4.1) is nonlinear, using iterative methods, the best guaranteed solution to (4.31) is a local optimum. More importantly, for the implementation on embedded systems with limited computational resources, the iteration number may be restricted for real-time control. In other words, the compatibility with suboptimal solutions introduces the flexibility between numerical efficiency and control performance. We can conveniently make the trade-off by specifying the maximum iteration number without destabilizing the tracking control.

4.3.2 Main Results

In this section, we first construct an auxiliary tracking controller using the Lyapunov-based backstepping technique and then analyze the recursive feasibility of the optimization problem and closed-loop stability of the LMPC under Algorithm 4.1.

To construct the contraction constraint in (4.31e), we need to find a state feedback controller together with the corresponding Lyapunov function. The Lyapunov-based nonlinear controller can be designed via the backstepping technique for the AUV trajectory tracking.

Define the following change of variables:

$$\dot{\boldsymbol{\eta}}_r = \dot{\boldsymbol{\eta}}_d - \tilde{\boldsymbol{\eta}}, \tag{4.32a}$$

$$\mathbf{v}_r = \mathbf{R}^{\mathrm{T}}(\psi)\dot{\boldsymbol{\eta}}_r, \tag{4.32b}$$

$$\mathbf{s} = \dot{\boldsymbol{\eta}} - \dot{\boldsymbol{\eta}}_r, \tag{4.32c}$$

where $\tilde{\boldsymbol{\eta}} = \boldsymbol{\eta} - \boldsymbol{\eta}_d$ is the position tracking error. Considering the kinematic equations (2.33), we have

$$\dot{\boldsymbol{\eta}} - \dot{\boldsymbol{\eta}}_d = \mathbf{R}(\psi)(\mathbf{v} - \mathbf{v}_d). \tag{4.33}$$

We view $\mathbf{v}$ as a virtual control that stabilizes the trajectory-tracking control

$$\mathbf{R}(\psi)\mathbf{v} = \mathbf{s} + \boldsymbol{\alpha}_1. \tag{4.34}$$

Choosing $\boldsymbol{\alpha}_1 = \dot{\boldsymbol{\eta}}_r$ and substituting it into (4.33)–(4.34) yields

$$\dot{\tilde{\boldsymbol{\eta}}} = \mathbf{s} + \boldsymbol{\alpha}_1 - \mathbf{R}(\psi)\mathbf{v}_d = -\tilde{\boldsymbol{\eta}} + \mathbf{s}. \tag{4.35}$$

Consider the following function:

$$V_1 = \frac{1}{2}\tilde{\boldsymbol{\eta}}^{\mathrm{T}}\mathbf{K}_p\tilde{\boldsymbol{\eta}}, \tag{4.36}$$

where $\mathbf{K}_p = \mathbf{K}_p^{\mathrm{T}} > 0$ is a specified control gain matrix. Then the time derivative of V_1 becomes:

$$\dot{V}_1 = \tilde{\boldsymbol{\eta}}^{\mathrm{T}}\mathbf{K}_p\dot{\tilde{\boldsymbol{\eta}}} = -\tilde{\boldsymbol{\eta}}^{\mathrm{T}}\mathbf{K}_p\tilde{\boldsymbol{\eta}} + \mathbf{s}^{\mathrm{T}}\mathbf{K}_p\tilde{\boldsymbol{\eta}}. \tag{4.37}$$

Further construct the Lyapunov function candidate

$$V_2 = \frac{1}{2}\mathbf{s}^{\mathrm{T}}\mathbf{M}^*(\psi)\mathbf{s} + V_1, \tag{4.38}$$

where $\mathbf{M}^*(\psi) = \mathbf{R}(\psi)\mathbf{M}\mathbf{R}^{\mathrm{T}}(\psi)$.

Taking time derivative of V_2 results in:

$$\dot{V}_2 = \mathbf{s}^{\mathrm{T}}\mathbf{M}^*(\psi)\dot{\mathbf{s}} + \tfrac{1}{2}\mathbf{s}^{\mathrm{T}}\dot{\mathbf{M}}^*(\psi)\mathbf{s} + \dot{V}_1. \tag{4.39}$$

Substituting the dynamic equations (2.31), we have

$$\begin{aligned}\dot{V}_2 = &-\mathbf{s}^{\mathrm{T}}[\mathbf{C}^*(\mathbf{v},\psi) + \mathbf{D}^*(\mathbf{v},\psi)]\mathbf{s}\\ &+\mathbf{s}^{\mathrm{T}}\mathbf{R}(\psi)[\boldsymbol{\tau} - \mathbf{M}\dot{\mathbf{v}}_r - \mathbf{C}(\mathbf{v})\mathbf{v}_r - \mathbf{D}(\mathbf{v})\mathbf{v}_r - \mathbf{g}(\boldsymbol{\eta})]\\ &+\tfrac{1}{2}\mathbf{s}^{\mathrm{T}}\dot{\mathbf{M}}^*(\psi)\mathbf{s} - \tilde{\boldsymbol{\eta}}^{\mathrm{T}}\mathbf{K}_p\tilde{\boldsymbol{\eta}} + \mathbf{s}^{\mathrm{T}}\mathbf{K}_p\tilde{\boldsymbol{\eta}},\end{aligned} \tag{4.40}$$

where $\mathbf{C}^*(\mathbf{v},\psi) = \mathbf{R}(\psi)[\mathbf{C}(\mathbf{v}) - \mathbf{M}\mathbf{R}^{\mathrm{T}}(\psi)\dot{\mathbf{R}}(\psi)]\mathbf{R}^{\mathrm{T}}(\psi)$ and $\mathbf{D}(\mathbf{v},\psi)^* = \mathbf{R}(\psi)\mathbf{D}(\mathbf{v})\mathbf{R}^{\mathrm{T}}(\psi)$.

From model property P-2, it can be verified that

$$\mathbf{s}^{\mathrm{T}}(\dot{\mathbf{M}}^*(\psi) - 2\mathbf{C}^*(\mathbf{v},\psi))\mathbf{s} = 0, \quad \forall\, \mathbf{v}, \psi, \mathbf{s}. \tag{4.41}$$

Therefore, if we choose the following control law:

$$\boldsymbol{\tau}(\mathbf{x}) = \mathbf{M}\dot{\mathbf{v}}_r + \mathbf{C}\mathbf{v}_r + \mathbf{D}\mathbf{v}_r + \mathbf{g} - \mathbf{R}^{\mathrm{T}}\mathbf{K}_p\tilde{\boldsymbol{\eta}} - \mathbf{R}^{\mathrm{T}}\mathbf{K}_d\mathbf{s}, \tag{4.42}$$

where $\mathbf{K}_d > 0$ is another user specified control gain matrix, and (4.40) becomes

$$\dot{V}_2 = -\mathbf{s}^{\mathrm{T}}[\mathbf{D}^*(\mathbf{v},\psi) + \mathbf{K}_d]\mathbf{s} - \tilde{\boldsymbol{\eta}}^{\mathrm{T}}\mathbf{K}_p\tilde{\boldsymbol{\eta}}. \tag{4.43}$$

From model property P-4, we have $\dot{V}_2 \leq 0$. Then by standard Lyapunov arguments, the closed-loop system under (4.42) is globally asymptotically stable with respect to the equilibrium $[\tilde{\boldsymbol{\eta}}, \mathbf{s}] = [\mathbf{0}, \mathbf{0}]$.

Therefore, the detailed expression of the contraction constraint (4.31e) corresponding to (4.42) is

$$\begin{aligned}&-\hat{\mathbf{s}}(0)^{\mathrm{T}}\mathbf{D}^*(\hat{\mathbf{v}}(0),\hat{\psi}(0))\hat{\mathbf{s}}(0) + \hat{\mathbf{s}}(0)^{\mathrm{T}}\mathbf{R}(\hat{\psi}(0))[\mathbf{B}\hat{\mathbf{u}}(0)\\ &\quad -\mathbf{M}\dot{\hat{\mathbf{v}}}_r(0) - \mathbf{C}(\hat{\mathbf{v}}(0))\hat{\mathbf{v}}_r(0) - \mathbf{D}(\hat{\mathbf{v}}(0))\hat{\mathbf{v}}_r(0)\\ &\quad -\mathbf{g}(\hat{\boldsymbol{\eta}}(0))] - \tilde{\boldsymbol{\eta}}(0)^{\mathrm{T}}\mathbf{K}_p\tilde{\boldsymbol{\eta}}(0) + \hat{\mathbf{s}}(0)^{\mathrm{T}}\mathbf{K}_p\tilde{\boldsymbol{\eta}}(0)\\ &\leq -\hat{\mathbf{s}}(0)^{\mathrm{T}}[\mathbf{D}^*(\hat{\mathbf{v}}(0),\hat{\psi}(0)) + \mathbf{K}_d]\hat{\mathbf{s}}(0) - \tilde{\boldsymbol{\eta}}(0)^{\mathrm{T}}\mathbf{K}_p\tilde{\boldsymbol{\eta}}(0).\end{aligned} \tag{4.44}$$

Note that we construct the contraction constraint with the help of the auxiliary nonlinear backstepping controller. However, in principle, any other Lyapunov-based tracking controller, such as sliding mode control and dynamic surface control, can be employed.

Now we analyze the recursive feasibility of the problem (4.31) and the closed-loop stability under Algorithm 4.1.

Lemma 4.1 *For the AUV system (4.1), suppose Assumption 4.3 is satisfied. If the backstepping control law (4.42) is applied to the AUV, then the Coriolis and centripetal matrix* $\mathbf{C}(\mathbf{v})$ *and damping matrix* $\mathbf{D}(\mathbf{v})$ *are such that the following relations* *hold:*

$$\|\mathbf{C}(\mathbf{v})\|_\infty \le \bar{c} = 2\sqrt{2}\bar{m}\bar{\eta}_{d1} + 4\sqrt{2}\bar{m}\|\gamma(0)\|_2, \tag{4.45}$$

$$\|\mathbf{D}(\mathbf{v})\|_\infty \le \bar{d} = \bar{d}_1 + \bar{d}_2(\sqrt{2}\bar{\eta}_{d1} + 2\sqrt{2}\|\gamma(0)\|_2), \tag{4.46}$$

where $\gamma(t) = \mathrm{col}(\tilde{\boldsymbol{\eta}}(t), \mathbf{s}(t))$, $\bar{d}_1 = \max\{|X_u|, |Y_v|, |N_r|\}$ *and* $\bar{d}_2 = \max\{D_u, D_v, D_r\}$.

Proof Define $\gamma' = \mathrm{col}(\tilde{\boldsymbol{\eta}}, \mathbf{R}^{\mathrm{T}}(\psi)\mathbf{s})$ and we can reformulate the Lyapunov function (4.38) as $V_2 = \frac{1}{2}\gamma'^{\mathrm{T}}\Pi\gamma'$ with $\Pi = \mathrm{diag}(\mathbf{K}_p, \mathbf{M})$. Since $\dot{V}_2 \le 0$, we have $\|\gamma'(t)\|_2 \le \|\gamma'(0)\|_2$. Furthermore, from the model property P-3 we have $\|\gamma'(t)\|_2 = \|\gamma(t)\|_2$ followed by

$$\|\gamma(t)\|_2 \le \|\gamma(0)\|_2. \tag{4.47}$$

By definition, we have $\|\tilde{\boldsymbol{\eta}}\|_\infty \le \|\gamma\|_\infty$ and $\|\mathbf{s}\|_\infty \le \|\gamma\|_\infty$. Then the following holds:

$$\|\dot{\tilde{\boldsymbol{\eta}}}\|_\infty = \|\mathbf{s} - \tilde{\boldsymbol{\eta}}\|_\infty \le \|\mathbf{s}\|_\infty + \|\tilde{\boldsymbol{\eta}}\|_\infty \le 2\|\gamma\|_\infty. \tag{4.48}$$

Considering that $\|\gamma\|_\infty \le \|\gamma\|_2$, we have

$$\|\dot{\tilde{\boldsymbol{\eta}}}(t)\|_\infty \le 2\|\gamma(0)\|_2. \tag{4.49}$$

Since $\|\dot{\boldsymbol{\eta}}\|_\infty = \|\dot{\boldsymbol{\eta}}_d + \dot{\tilde{\boldsymbol{\eta}}}\|_\infty \le \|\dot{\boldsymbol{\eta}}_d\|_\infty + \|\dot{\tilde{\boldsymbol{\eta}}}\|_\infty$, together with (4.49) and Proposition 4.2, it follows that

$$\|\dot{\boldsymbol{\eta}}(t)\|_\infty \le \bar{\eta}_{d1} + 2\|\gamma(0)\|_2. \tag{4.50}$$

By the kinematic equations (2.33) and the model property P-3, we have $\|\mathbf{v}\|_\infty = \|\mathbf{R}^{\mathrm{T}}(\psi)\dot{\boldsymbol{\eta}}\|_\infty \le \sqrt{2}\|\dot{\boldsymbol{\eta}}\|_\infty$ due to the fact that $\|\mathbf{R}^{\mathrm{T}}(\psi)\|_\infty = \max\{|\cos\psi| + |\sin\psi|, 1\} \le \sqrt{2}$. Therefore, it holds that

$$\|\mathbf{v}(t)\|_\infty \le \sqrt{2}\bar{\eta}_{d1} + 2\sqrt{2}\|\gamma(0)\|_2. \tag{4.51}$$

Having the model property P-1 and taking infinity norm on (2.30) yields

$$\|\mathbf{C}(\mathbf{v})\|_\infty \le |M_{\dot{u}}u| + |M_{\dot{v}}v| \le \bar{m}(|u| + |v|). \tag{4.52}$$

Further considering (4.51), we have

$$\|\mathbf{C}(\mathbf{v})\|_\infty \le 2\sqrt{2}\bar{m}\bar{\eta}_{d1} + 4\sqrt{2}\bar{m}\|\gamma(0)\|_2. \tag{4.53}$$

From the detailed damping matrix expression (2.29), by definition, we have $\|\mathbf{D}(\mathbf{v})\|_\infty = \max\{|X_u + D_u|u||, |Y_v + D_v|v||, |N_r + D_r|r||\}$. Further considering (4.51), it yields

$$\|\mathbf{D}(\mathbf{v})\|_\infty \le \bar{d}_1 + \bar{d}_2(\sqrt{2}\bar{\eta}_{d1} + 2\sqrt{2}\|\gamma(0)\|_2). \tag{4.54}$$

□

Theorem 4.3 *Suppose the control gain matrices are chosen as* $\mathbf{K}_p = \mathrm{diag}(k_{pi})$ *and* $\mathbf{K}_d = \mathrm{diag}(k_{di})$*, positive definite. Let* $\bar{k}_p = \max\{k_{pi}\}$ *denote the largest element in* $\mathbf{K}_p$ *and* $\bar{k}_d = \max\{k_{di}\}$ *denote the largest element in* $\mathbf{K}_d$*. Suppose Assumptions 4.1 and 4.3 can hold. For* $h(\mathbf{x}) = \mathbf{B}^{+}\boldsymbol{\tau}(\mathbf{x})$*, if the following relation can be satisfied:*

$$\bar{m}\bar{v}_{r1} + (\bar{c} + \bar{d})\bar{v}_r + \bar{g} + \sqrt{2}(\bar{k}_p + \bar{k}_d)\|\gamma(0)\|_2 \leq \tau_{\max}, \tag{4.55}$$

with $\bar{v}_{r1} = \sqrt{2}\bar{\eta}_{d2} + 2\bar{\eta}_{d1}^2 + (2\sqrt{2} + 6\bar{\eta}_{d1})\|\gamma(0)\|_2 + 4\|\gamma(0)\|_2^2$ *and* $\bar{v}_r = \sqrt{2}(\bar{\eta}_{d1} + \|\gamma(0)\|_2)$*, and* τ_{max} *follows (4.11), then the LMPC (4.31) admits recursive feasibility, i.e.,* $|h(\hat{\mathbf{x}}(t))| \leq \mathbf{u}_{\max}$ *for all* $t \geq 0$ *where* $\mathbf{u}_{\max} = u_{\max}\mathbf{1}$.

Proof We notice that given the current system state $\mathbf{x}(t)$, $h(\hat{\mathbf{x}})$ is always feasible for the LMPC problem (4.31) if $|h(\hat{\mathbf{x}})| \leq \mathbf{u}_{\max}$ can be satisfied.

Since we have Proposition 4.2 and the fact that $\|\hat{\tilde{\boldsymbol{\eta}}}\|_\infty \leq \|\hat{\gamma}\|_\infty \leq \|\gamma(0)\|_2$, it can be verified that $\|\hat{\mathbf{v}}_r\|_\infty \leq \bar{v}_r$ with $\bar{v}_r = \sqrt{2}(\bar{\eta}_{d1} + \|\gamma(0)\|_2)$ by taking infinity norm on both sides of (4.32).

Taking time derivative on (4.32a) and on (4.32a) we have

$$\begin{aligned} \dot{\mathbf{v}}_r &= \mathbf{R}^{\mathrm{T}}(\psi)\ddot{\boldsymbol{\eta}}_r - \dot{\psi}\boldsymbol{\Omega}(\psi)\dot{\boldsymbol{\eta}}_r \\ &= \mathbf{R}^{\mathrm{T}}(\psi)(\ddot{\boldsymbol{\eta}}_d - \dot{\tilde{\boldsymbol{\eta}}}) - r\boldsymbol{\Omega}(\psi)(\dot{\boldsymbol{\eta}}_d - \tilde{\boldsymbol{\eta}}), \end{aligned} \tag{4.56}$$

where

$$\boldsymbol{\Omega}(\psi) = \begin{bmatrix} \sin\psi & -\cos\psi & 0 \\ \cos\psi & \sin\psi & 0 \\ 0 & 0 & 0 \end{bmatrix}, \tag{4.57}$$

which admits $\|\boldsymbol{\Omega}(\psi)\|_\infty \leq \sqrt{2}$. Then taking infinity norm on (4.56) and combining (4.49) and (4.51), we have $\|\dot{\hat{\mathbf{v}}}_r\|_\infty \leq \bar{v}_{r1}$ with $\bar{v}_{r1} = \sqrt{2}\bar{\eta}_{d2} + 2\bar{\eta}_{d1}^2 + (2\sqrt{2} + 6\bar{\eta}_{d1})\|\gamma(0)\|_2 + 4\|\gamma(0)\|_2^2$.

Then taking infinity norm on both sides of (4.42), together with Lemma 4.1 and the model property P-6, we have

$$\|\boldsymbol{\tau}(\hat{\mathbf{x}})\|_\infty \leq \bar{m}\bar{v}_{r1} + (\bar{c} + \bar{d})\bar{v}_r + \bar{g} + \sqrt{2}(\bar{k}_p + \bar{k}_d)\|\gamma(0)\|_2. \tag{4.58}$$

If (4.55) can be satisfied, then $\|\boldsymbol{\tau}(\hat{\mathbf{x}})\|_\infty \leq \tau_{\max}$ can hold. With (4.11), we can guarantee that $\|h(\hat{\mathbf{x}}(s))\|_\infty \leq u_{\max}$ will be always satisfied. This completes the proof. □

Note that the nonlinear backstepping controller acts as an initial guess for the optimization problem (4.31) but is never applied to the AUV system. As a result, the feasibility is guaranteed by the auxiliary controller and the control performance will be optimized by the LMPC method.

Theorem 4.4 *Suppose that Assumptions 4.1 and 4.3 hold, then the closed-loop system under Algorithm 4.1 is asymptotically stable with respect to the equilibrium* $[\tilde{\eta}, \mathbf{s}] = [\mathbf{0}, \mathbf{0}]$, *i.e., the AUV converges to the desired trajectory* $p(t)$ *with the LMPC-based trajectory-tracking control.*

Proof According to the Lyapunov function $V_2(\mathbf{x})$ in (4.38), continuously differentiable and radically unbounded, by converse Lyapunov theorems (Khalil 1996), there exist functions $\beta_i(\cdot)$, $i = 1, 2, 3$ which belong to class $\mathcal{K}_\infty$ such that the following inequalities hold:

$$\beta_1(\|\mathbf{x}\|) \leq V_2(\mathbf{x}) \leq \beta_2(\|\mathbf{x}\|), \tag{4.59a}$$

$$\frac{\partial V}{\partial \mathbf{x}} \mathbf{f}(\mathbf{x}, h(\mathbf{x})) \leq -\beta_3(\|\mathbf{x}\|). \tag{4.59b}$$

Considering the contraction constraint (4.31e) and that the optimal solution $\boldsymbol{\kappa}(s)$ will be implemented for one sampling period, we have

$$\frac{\partial V}{\partial \mathbf{x}} \mathbf{f}(\mathbf{x}, \mathbf{u}(\mathbf{x})) \leq \frac{\partial V}{\partial \mathbf{x}} \mathbf{f}(\mathbf{x}, h(\mathbf{x})) \leq -\beta_3(\|\mathbf{x}\|). \tag{4.60}$$

By standard Lyapunov arguments we claim that the closed-loop system under Algorithm 4.1 is asymptotically stable with a guaranteed region of attraction

$$\mathcal{X} = \{\mathbf{x} \in \mathbb{R}^n \mid (4.55)\}. \tag{4.61}$$

Furthermore, $\mathcal{X}$ can be enlarged by shrinking the magnitude of the control gains $\bar{k}_p$ and $\bar{k}_d$. □

Remark 4.3 Although the asymptotic stability relies only on the positive definiteness of the control gain matrices $\mathbf{K}_p$ and $\mathbf{K}_d$, the tracking control performance with the backstepping controller $\boldsymbol{\tau}(\mathbf{x})$ in (4.42) is determined by the magnitude of the control gains. From (4.43), we can see that smaller values of $\bar{k}_p$ and $\bar{k}_d$ result in slower convergence. However, for the proposed LMPC-based trajectory-tracking control, thanks to the optimization procedure, the controller can automatically make full use of the thrust capability to generate the best possible tracking control with respect to the objective function (4.31a) even if we have selected small control gains for a large region of attraction.

4.3.3 Simulation Study

This section presents simulation results of the AUV tracking control, which highlight the advantages of the proposed LMPC method. All the simulations are based on the experimentally identified dynamic model of Falcon. The model parameter details can be found in Table 2.1.

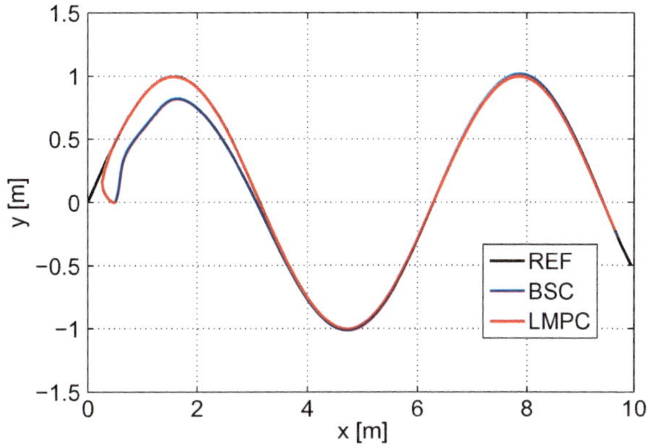

Fig. 4.7 AUV trajectory in local level plane—Case I

Two desired trajectories are used to test the AUV tracking control. The first one (Case I) is a sinusoidal trajectory defined as follows:

$$p(t) = \begin{cases} x_d = 0.5t, \\ y_d = \sin(0.5t), \end{cases} \tag{4.62}$$

and the second one (Case II) is an eight-shaped trajectory defined by

$$p(t) = \begin{cases} x_d = -\sin(0.5t), \\ y_d = \sin(0.25t). \end{cases} \tag{4.63}$$

For the LMPC tracking controller, the following parameters are used: The sampling period is $\delta = 0.1$ s; the prediction horizon is $T = 5\delta$; the weighting matrices are chosen as $Q = \text{diag}(10^5, 10^5, 10^3, 10^2, 10^2, 10^2)$, $R = \text{diag}(10^{-4}, 10^{-4}, 10^{-4}, 10^{-4})$, and $P = \text{diag}(10^3, 10^3, 10^2, 10, 10, 10)$; and the limit on each thruster is 500 [N]. The control gains $\mathbf{K}_p = \mathbf{K}_d = \text{diag}(1, 1, 1)$. The initial condition is $\mathbf{x}(0) = [0.5, 0, 0, 0, 0, 0]^{\mathrm{T}}$.

To solve the LMPC problem (4.31) numerically, we discretize the problem and then solve the corresponding Karush–Kuhn–Tucker (KKT) conditions by the SQP method (Antoniou and Lu 2007).

The trajectory-tracking results for Case I are shown in Figs. 4.7 and 4.8. The blue curve is the simulated AUV trajectories using backstepping control (denoted by BSC), and the red curve is the AUV trajectories with the proposed LMPC control, while the black curve is the desired sinusoidal trajectory. It can be observed that both backstepping control and LMPC drive the vehicle to the desired trajectory, which verifies the closed-loop stability. But obviously the LMPC controller generates a much faster convergence than the backstepping controller. This is because we have selected small control gain matrices $\mathbf{K}_p$ and $\mathbf{K}_d$ for a large region of attraction. The simulation results demonstrate the enhanced tracking control performance brought by the online optimization.

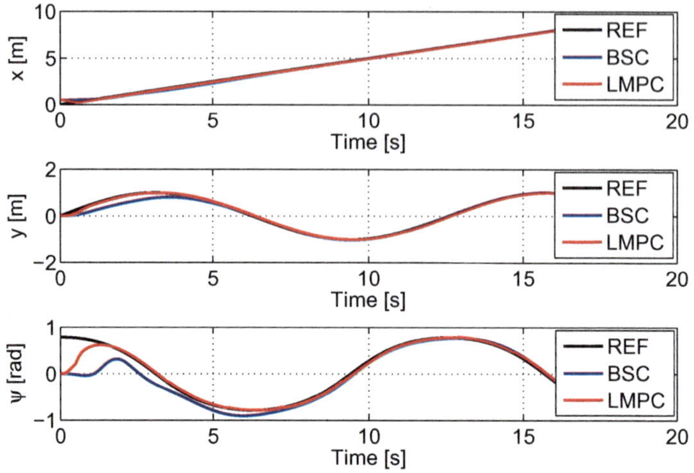

Fig. 4.8 State trajectories—Case I

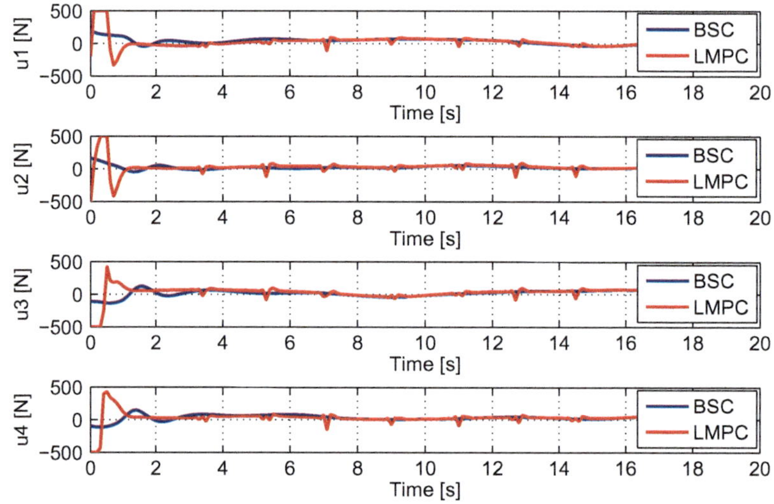

Fig. 4.9 Control input signals—Case I

The required control forces for each thruster are plotted in Fig. 4.9. At the beginning of the tracking, we observe that the LMPC controller fully uses the onboard thrust capability to generate the fastest possible convergence while respecting the physical limit of thrusters. The magnitude of control commands stays within the permitted range as expected.

The simulation results for Case II are provided through Figs. 4.10, 4.11, and 4.12. Similar observations can be made: The AUV converges faster to the desired trajectory

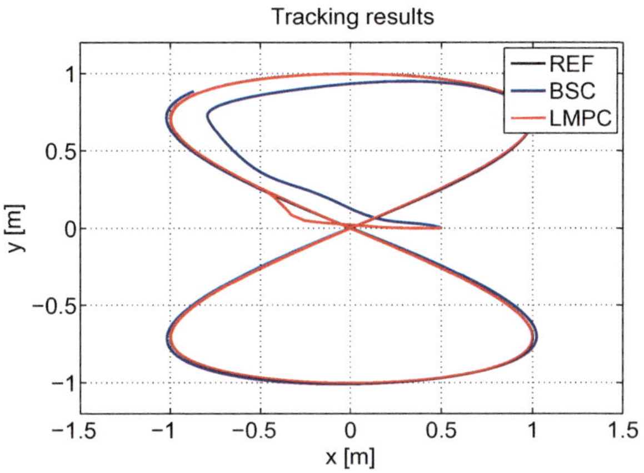

Fig. 4.10 AUV trajectory in local level plane—Case II

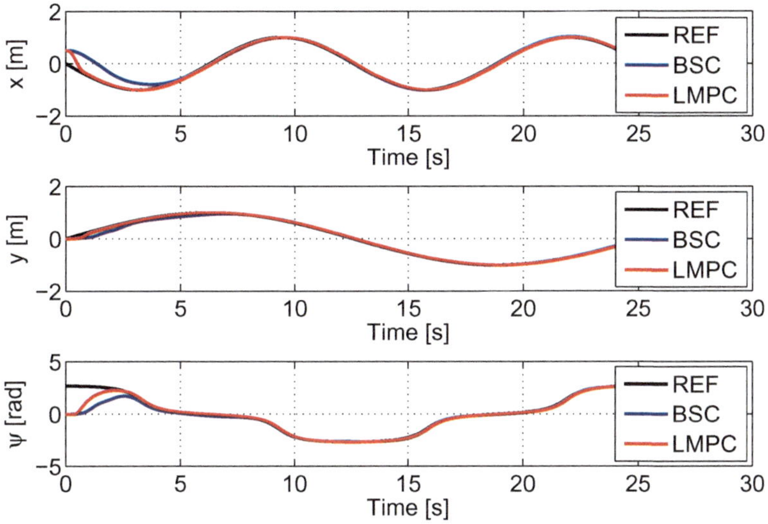

Fig. 4.11 State trajectories—Case II

when adopting the proposed LMPC-based tracking control and the calculated control commands are always feasible for the real system.

To test the robustness of the backstepping and LMPC controllers, we simulate the AUV tracking control under the following test condition: There exist 30% model parameter error and an ocean current introduced disturbance $\mathbf{w} = [100\text{N}, 100\text{N}, 0\text{Nm}]^{\text{T}}$. Note that this is a very strict test condition, and it is more of a test than introducing some obscure coupling of vehicle DOFs that may or may not exist on a real AUV.

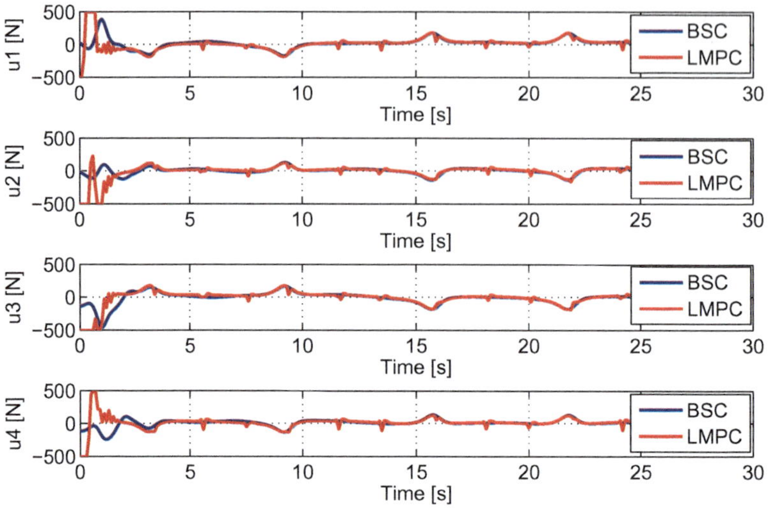

Fig. 4.12 Control input signals—Case II

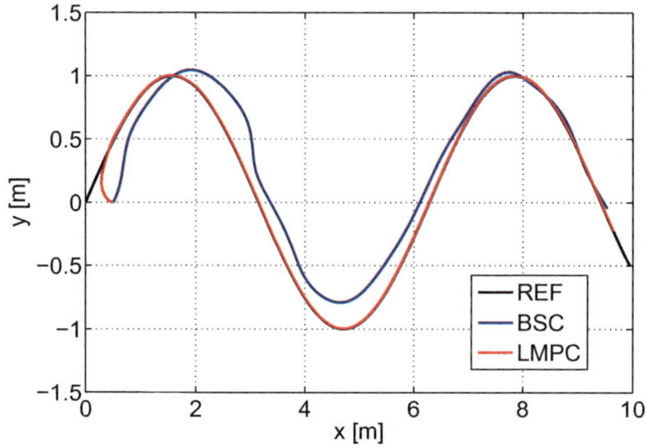

Fig. 4.13 AUV trajectory in local level plane (with disturbance)—Case I

From the simulation results illustrated in Figs. 4.13, 4.14, 4.15, and 4.16, we find that the LMPC-based tracking control still successfully leads the AUV well converged to the desired trajectory, while the tracking control with backstepping technique suffers from large tracking errors. The quantitative comparisons for both cases are summarized in Tables 4.1 and 4.2, where tremendous improvement in mean square errors (MSEs) can be witnessed. Roughly speaking, the MSEs are more than 20 times smaller than the LMPC, especially for Case II. This is because LMPC can leverage online optimization to schedule an appropriate control gain to compensate for the disturbances well, whereas backstepping method is a fixed gain controller and lacks such flexibility. The LMPC, therefore, is robust to model uncertainties and

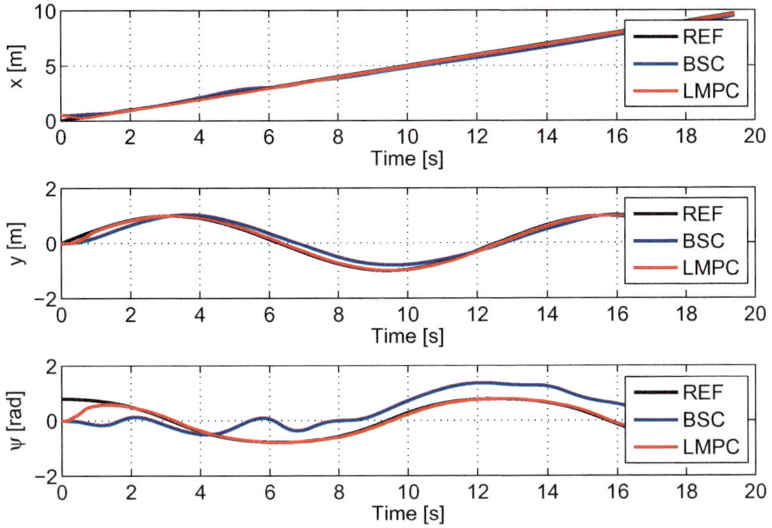

Fig. 4.14 State trajectories (with disturbance)—Case I

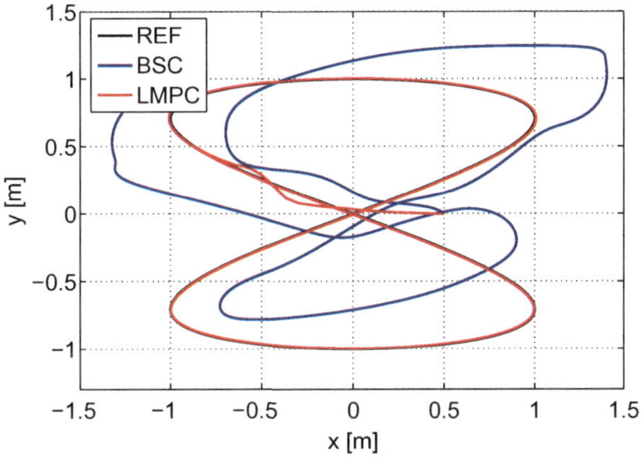

Fig. 4.15 AUV trajectory in local level plane (with disturbance)—Case II

external disturbances, which exhibits great potential for the motion control of general marine vessels.

4.4 Notes and Summary

In this chapter, we have proposed a novel LMPC framework to solve the AUV motion control problems. The LMPC presents a convenient and effective tool to allocate onboard computing resources to improve the performance of the existing motion

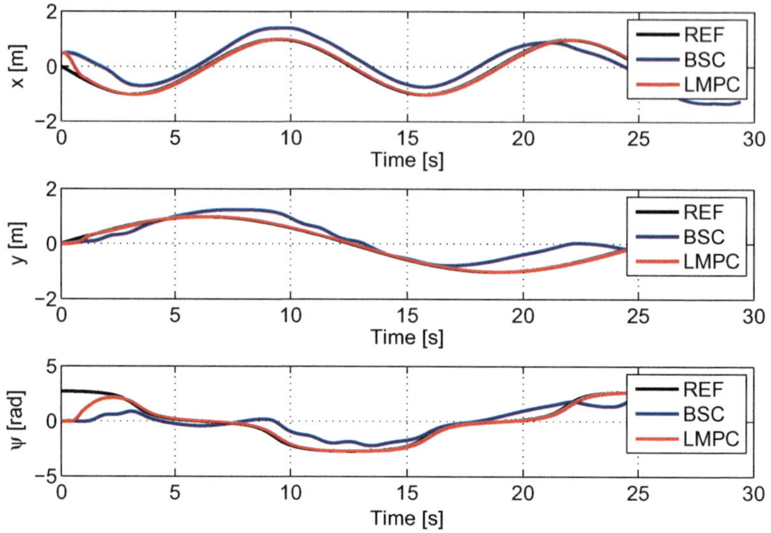

Fig. 4.16 State trajectories (with disturbance)—Case II

Table 4.1 MSEs for AUV tracking with disturbances—Case I

MSEs	Backstepping	LMPC	Improvement (%)
x (m^2)	0.0418	0.0052	87.5
y (m^2)	0.0297	0.0015	94.9
ψ (rad^2)	0.3207	0.0184	94.3

Table 4.2 MSEs for AUV tracking with disturbances—Case II

MSE	Backstepping	LMPC	Improvement (%)
x (m^2)	0.1483	0.0059	96.0
y (m^2)	0.0977	0.0005	99.5
ψ (rad^2)	1.0081	0.2480	75.4

control system. Two fundamental AUV motion control problems, namely dynamic positioning and trajectory tracking, were investigated under the LMPC framework and ended up with the controller design.

Using the nonlinear PD control law as the base DP controller, the conditions for recursive feasibility and closed-loop stability of the LMPC control system were derived. A quasi-global closed-loop stability property was claimed for the LMPC-based DP control: The ROA can be specified as arbitrarily large. Also, built on a nonlinear backstepping tracking control law, the LMPC-based trajectory-tracking controller was designed with guaranteed feasibility and stability. A guaranteed ROA was characterized explicitly.

Simulation studies on the dynamic model of Falcon suggested that the dynamic positioning and trajectory-tracking performance could be dramatically improved by the online optimization incorporated in the LMPC control. Meanwhile, through the robustness test, the superior robustness of the LMPC control system was demonstrated, which indicated that the proposed LMPC method was very promising and could be applied to real AUVs (large model mismatch) under real working conditions (large disturbances).

References

Antoniou A, Lu WS (2007) Practical optimization. Springer

Christofides PD, Liu J, De La Pena DM (2011) Networked and distributed predictive control: methods and nonlinear process network applications. Springer

Fossen TI (2002) Marine control systems: guidance, navigation, and control of ships, rigs and underwater vehicles. Marine Cybernetics, Trondheim, Norway

Fossen TI, Grovlen A (1998) Nonlinear output feedback control of dynamically positioned ships using vectorial observer backstepping. IEEE Trans Control Syst Technol 6(1):121–128

Katebi MR, Grimble MJ, Zhang Y (1997) H_∞ robust control design for dynamic ship positioning. IEE Proc Control Theory Appl 144(2):110–120

Khalil HK (1996) Nonlinear systems. Prentice Hall, New York

Li H, Shi Y (2017) Robust receding horizon control for networked and distributed nonlinear systems. Springer

Mayne DQ (2014) Model predictive control: recent developments and future promise. Automatica 50(12):2967–2986

Mayne DQ, Rawlings JB, Rao CV, Scokaert PO (2000) Constrained model predictive control: stability and optimality. Automatica 36(6):789–814

Mhaskar P, El-Farra NH, Christofides PD (2006) Stabilization of nonlinear systems with state and control constraints using Lyapunov-based predictive control. Syst Control Lett 55(8):650–659

Pannocchia G, Rawlings JB, Wright SJ (2011) Inherently robust suboptimal nonlinear MPC: theory and application. In: Proceedings of 2011 50th IEEE conference on decision and control and European control conference, pp 3398–3403. IEEE

Proctor AA (2014) Semi-autonomous guidance and control of a Saab SeaEye Falcon ROV. Ph.D thesis

Rawlings JB, Mayne DQ, Diehl M (2017) Theory, computation, and design. Nob Hill Publishing Madison

Sørensen AJ (2011) A survey of dynamic positioning control systems. Ann Rev Control 35(1):123–136

Sørensen AJ, Sagatun SI, Fossen TI (1996) Design of a dynamic positioning system using model-based control. Control Eng Pract 4(3):359–368

Tannuri EA, Agostinho AC, Morishita HM, Moratelli L Jr (2010) Dynamic positioning systems: an experimental analysis of sliding mode control. Control Eng Pract 18(10):1121–1132

Veksler A, Johansen TA, Borrelli F, Realfsen B (2016) Dynamic positioning with model predictive control. IEEE Trans Control Syst Technol 24(4):1340–1353

Chapter 5
Multi-objective Model Predictive Control for Path-Following Control of an AUV

5.1 Introduction

5.1.1 Research Background and Contributions

In the previous chapters, we have studied the trajectory-tracking problem of the AUV. In trajectory-tracking problems, the reference trajectory defines the desired footprints with precise time stamps, i.e., it specifies when should the vehicle be where. However, in many circumstances, we may be interested in following a specific geometric pattern in the workspace, and the temporal requirement can be relaxed, i.e., we mainly care about the footprints but not the exact time the vehicle gets there. If this is the case, we are considering the PF problem.

The PF control problem has been extensively studied in the past several decades. The work reported in (Micaelli and Samson 1993) presents a powerful PF control structure for land robots. The control structure, however, exploits the kinematic characteristics only. The techniques therein cannot be directly applied to AUVs. In (Encarnacao and Pascoal 2003), Encarnacao et al. took advantage of the backstepping technique and recursively designed the PF controller for the underwater vehicle. In (Encarnação et al. 2000), Encarnacao et al. further considered more practical scenarios and extended the PF controller design to compensate the ocean currents. However, the controllers in (Encarnacao and Pascoal 2003; Encarnação et al. 2000) are designed based on that the target point is selected to be the closest point on the path relative to the vehicle; hence, these methods share an inherent limitation: There exist singularities for certain initial positions of the vehicle, located exactly at the centers of path curvature. To solve this serious problem, in (Lapierre and Soetanto 2007) Lapierre et al. proposed a novel PF control strategy. The target point was changed to a virtual particle moving along the path and possessing independent dynamic properties. The novel strategy essentially created an extra DOF which could be exploited to achieve additional specifications such as a desired speed assignment. Following this strategy, in (Lapierre and Jouvencel 2008) Lapierre et al. further developed the

Y. Shi et al., *Advanced Model Predictive Control for Autonomous Marine Vehicles*, Advances in Industrial Control, https://doi.org/10.1007/978-3-031-19354-5_5

robust PF controller by involving an adaptive law which well handles the parametric uncertainty in the AUV model. Unfortunately, the preceding PF controller designs have no capability to incorporate the system constraints. Accommodating such constraints has been the focus of recent studies of MPC for PF problems (Faulwasser et al. 2009; Faulwasser and Findeisen 2010; Böck and Kugi 2013).

The MPC provides a flexible framework that accommodates the complicated system dynamics and nonlinear system constraints. These features make MPC desirable for a variety of control problems (Mayne 2014; Qin and Badgwell 2000). In the context of path-following control, there exist several pioneering works in the literature. In (Pavlov et al. 2009), the PF problem was studied for marine vessels, in which MPC was used to optimize the lookahead distance in the LOS guidance law. In (Oh and Sun 2010), a complete linear MPC structure was developed for the LOS-based PF control rather than just adding an optimization of the lookahead parameter. However, the LOS guidance law is only valid for PF of line segments (waypoint routes). To solve the PF problem with a desired curved path, Faulwasser et al. (Faulwasser et al. 2009) provided the standard MPC formulation with an augmented system consisting of the vehicle dynamics and the path dynamics. Sufficient stabilizing conditions were derived. An extended work in terms of output MPC version was presented in (Faulwasser and Findeisen 2010). The PF control designs in (Faulwasser et al. 2009; Faulwasser and Findeisen 2010) only guarantee a nonzero forward motion and cannot easily accommodate a desired speed assignment. Therefore, the incorporation of the speed assignment into the MPC formulation was studied in (Alessandretti et al. 2013) for underactuated vehicles. In (Yu et al. 2015) Yu et al. took another approach that transforms the PF problem into a regulation problem where the initial state of the path dynamics was also viewed as a decision variable. The PF task can also be accomplished with a trajectory-tracking controller by explicitly assigning a timing law to the path dynamics. In this way, however, it loses the flexibility of adjusting the velocity of the vehicle for possible better path-following performance. In (Shen et al. 2017), the coupled path planning and tracking control problem was solved with a model predictive trajectory-tracking control formulation. The MPC-based PF control applied to industrial robots and tower cranes can be found in (Faulwasser et al. 2016) and (Böck and Kugi 2013), respectively.

In practice, however, the PF control always prioritizes the path convergence requirement over the speed assignment; therefore, the MPC framework proposed in (Alessandretti et al. 2013) cannot provide flexibility to adjust the relative priorities of the path convergence and speed assignment objectives. To systematically address the practical priority issue among several expected objectives, it is natural to resort to the MOMPC method. The MOMPC has been receiving increasing attentions from the system control community (Bemporad and de la Peña 2009; Zavala et al. 2012; Kerrigan and Maciejowski 2002), because many practical applications involve meaningful multiple control objectives. Early results in MOMPC mainly focus on the problem modeling, i.e., how to formulate the interested control problem into the MOMPC framework. Examples include the sewer network application (Ocampo-Martinez et al. 2007) and the solar cooling plant application (Zambrano and Camacho 2002). In recent years, the MOMPC studies begin to emphasize the closed-loop properties such as stability and optimality. In (Bemporad and de la Peña 2009), Bemporad et al., adopted the WS method to solve the multi-objective optimization problem and ensured the closed-loop stability for a special class of linear systems by adding an intermediate procedure which determines appropriate weights. In (Zavala et al. 2012), the utopia-tracking strategy was employed for the MOMPC problem. By constructing the partial Lagrange function and under assumptions of strong duality and Lipschitz continuity, the utopia-tracking cost can be exploited to establish the nonincreasing property, which guarantees the closed-loop stability. The LO method which is used to solve the multi-objective optimization was studied for the MOMPC problem in (He et al. 2015). The optimal value function for the most important objective can be shown a valid Lyapunov function and thus guarantees the closed-loop stability. An interesting variant in terms of switched MPC was proposed in (Müller and Allgöwer 2012). The average dwell time was imposed between switches so that the cost function can be monotonically decreasing no matter how arbitrary the switches will be made.

The MOMPC has an enlarged capacity and exhibits an attractive framework to solve AUV path-following control problem. In this chapter, we propose a novel MOMPC formulation for the PF problem of an AUV so that the PF requirement prioritization can be explicitly incorporated. Regarding the competing objectives, the WS method and the LO methods will be investigated. For the WS-MOMPC method, a logistic function is introduced to automate the weight selection. To implement the WS-MOMPC PF control, the PMP is applied. For the LO-MOMPC method, the controller design that guarantees the path convergence is provided.

The main contributions of this chapter include the following:

- A novel MOMPC framework is developed for the AUV PF problem to explicitly incorporate the objective prioritization in the PF task.
- The WS and LO methods are proposed to solve the MOMPC problem; the implicit relationship between two methods is explored.
- A novel logistic function is introduced to deal with the weight selection associated with the WS-MOMPC method. With several properly defined barrier functions, the PMP is applied to implement the WS-MOMPC algorithm.
- Sufficient conditions that guarantee the convergence of the MOMPC-based PF control are derived; and a feasible solution to satisfy the conditions is provided for the LO-MOMPC implementation.

5.1.2 Chapter Organization

The remaining part of this chapter is organized as follows: In Sect. 5.2, the detailed problem formulation is presented. Two methods that solve the MOMPC problem are proposed in Sect. 5.3. In Sect. 5.4, the conditions that guarantee the convergence of the solution are explicitly derived. Section 5.5 presents the simulation results of the proposed MOMPC PF control, and some investigation into robustness of the MOMPC method is provided. Section 5.6 concludes the entire chapter.

The notation is explained as follows: The column operation $[\rho_1^{\mathrm{T}}, \ldots, \rho_n^{\mathrm{T}}]^{\mathrm{T}}$ is written as $\mathrm{col}(\rho_1, \ldots, \rho_n)$; diagonal operation is abbreviated by $\mathrm{diag}(\cdot)$; and the weighted norm $\sqrt{\rho^{\mathrm{T}} Q \rho}$ is denoted by $\|\rho\|_Q$. For a function $p(s)$, the derivative regarding s is represented by $p'(s)$ while the time derivative is denoted by $\dot{p}(s)$.

5.2 Problem Formulation

The AUV model studied for the PF control problem is established using the kinematic and the dynamic equations:

$$\dot{\mathbf{x}} = \begin{bmatrix} \mathbf{R}(\psi)\mathbf{v} \\ \mathbf{M}^{-1}(\mathbf{u} - \mathbf{C}(\mathbf{v})\mathbf{v} - \mathbf{D}(\mathbf{v})\mathbf{v} - \mathbf{g}(\eta)) \end{bmatrix} = f(\mathbf{x}, \mathbf{u}), \tag{5.1}$$

where the state vector $\mathbf{x} = [x, y, \psi, u, v, r]^{\mathrm{T}}$ is consisted of the pose and velocity of the vehicle, and the control vector $\mathbf{u} = [F_u, F_v, F_r]^{\mathrm{T}}$ is generalized thrust forces and moments. The detailed expression can be found in (2.32) and (2.34).

To facilitate the following mathematical derivation, we will directly use the discretized version of (5.1) in the MOMPC problem formulation

$$\mathbf{x}_{k+1} = f_d(\mathbf{x}_k, \mathbf{u}_k). \tag{5.2}$$

Note that the detailed expression of (5.2) is determined by the detailed discretization method. For a small sampling period, the Euler method or the Runge–Kutta method is practically sufficient.

5.2.1 *The Path-Following Problem*

Consider the following mapping:

$$s \mapsto p(s) \in \mathbb{R}^d, \; s \in [s_0, s_1] \subset \mathbb{R}, \tag{5.3}$$

which describes the geometric path

$$\mathcal{P} = \{\bar{p} \in \mathbb{R}^d \mid \bar{p} = p(s), s \in [s_0, s_1]\}, \tag{5.4}$$

in a d-dimensional output space. The scalar variable s is referred to as the path parameter and the domain should be closed but may be unbounded, i.e., $s_1 = +\infty$. The mapping p is assumed to be smooth and bounded. To facilitate the PF problem formulation, the path $\mathcal{P}$ can be interpreted as the output of the path dynamics governed by

$$\dot{s} = g(s, v_s), \ \ \bar{p} = p(s), \tag{5.5}$$

where v_s is the control input for the path dynamics.

Alternatively, the desired path $\mathcal{P}$ might also be defined in the n-dimensional state space by the mapping

$$s \mapsto \bar{p}_{\mathbf{x}}(s) \in \mathbb{R}^n, \ \ s \in [s_0, s_1] \subset \mathbb{R}, \tag{5.6}$$

in which the components of $\bar{p}_{\mathbf{x}}(s)$ are coupled according to the dynamics of the controlled system, hence cannot be freely chosen.

More precisely, the output path (5.4) defines a zero-path-error (ZPE) manifold (Faulwasser 2013) in the state space, described by the following mapping:

$$s \mapsto p_{\mathbf{x}}(s, \dot{s}, \ddot{s}, \ldots) \in \mathbb{R}^n, \ \ s \in [s_0, s_1] \subset \mathbb{R}. \tag{5.7}$$

We will particularly exploit the ZPE in the subsequent MOMPC formulation.

Then the PF problem that under study in this chapter can be formulated as determining the control signal $\mathbf{u}(t)$ such that the following requirements are met:

- Path convergence: The AUV system states converge to the ZPE manifold:

$$\lim_{t \to \infty} \|\mathbf{x}(t) - p_{\mathbf{x}}(s(t), \dot{s}(t), \ldots)\| = 0. \tag{5.8}$$

- Forward motion: The AUV follows the path in the direction of $\dot{s} \geq 0$.
- Speed assignment: When moving along the path, the desired speed profile is pursued: $\dot{s} \rightarrow \dot{s}_r$.
- Constraint satisfaction: The AUV system constraints such as thrust limits are always respected.

Note that the path convergence should be identified as the primary task and need to be guaranteed, and the speed assignment is the secondary task which can be sacrificed at times, in lieu of better performance on the primary one.

5.2.2 *Zero-Path-Error Manifold*

In general, finding an explicit parametrization of the ZPE manifold (5.7) could be difficult. Fortunately, for the AUV system, each state has its physical meaning, which provides a guideline for the parametrization.

Consider a desired path defined in the output space

$$\mathcal{P} = \{\bar{p} \in \mathbb{R}^2 \mid \bar{p} = [p_x(s), p_y(s)]^{\mathrm{T}}, \ s \in [s_0, s_1]\}. \tag{5.9}$$

Regarding the kinematic equations (2.34), we choose the AUV surge velocity u that is tangent to the desired path, then the ZPE manifold can be explicitly parameterized in the following way:

$$p_{\mathbf{x}}(s, \dot{s}) = [p_x(s), p_y(s), p_\psi(s), p_u(s, \dot{s}), p_v(s), p_r(s, \dot{s})]^{\mathrm{T}}, \tag{5.10}$$

with

$$p_\psi(s) = \mathrm{atan2}(p'_y, p'_x), \tag{5.11a}$$

$$p_u(s, \dot{s}) = \sqrt{p'^2_x + p'^2_y}\dot{s}, \tag{5.11b}$$

$$p_v(s) = 0, \tag{5.11c}$$

$$p_r(s, \dot{s}) = \frac{p'_x p''_y - p'_y p''_x}{p'^2_x + p'^2_y}\dot{s}. \tag{5.11d}$$

Note that since we have additionally confined the velocity of the AUV, (5.10)–(5.11) do not cover the entire ZPE manifold defined by the output path (5.9) but define a subset of the manifold which is convenient to parameterize. It specifies the unique reference for each state of the AUV to track, which facilitates the following convergence analysis.

For the manifold (5.10)–(5.11), only the first-order time derivative of the path parameter is needed. Therefore, we can choose the single integrator model to describe the path dynamics

$$\dot{s} = g(s, v_s) = v_s. \tag{5.12}$$

This choice may facilitate the implementation of the MOMPC algorithm in the sense that the forward motion requirement will be formulated as inequality constraints on the input variables rather than on the state variables. Then those indirect methods which are based on solving the PMP (Diehl et al. 2009; Graichen 2012) can be applied.

5.2.3 *The MOMPC Formulation*

Basically, the AUV PF problem considers two aspects: path convergence and speed assignment. The path convergence requires the AUV to converge to $\mathcal{P}$ as fast as possible, while the speed assignment requires the path parameter to move in a preferred pace. Since $\dot{s}_r$ is known, stringent fulfillment of the speed assignment results in the determined reference states $p_{\mathbf{x}}(s_r, \dot{s}_r)$ at each time instant. This substantially degrades the PF problem to a trajectory-tracking problem where the flexibility of adjusting forward speed may be lost.

To maintain this flexibility, the MOMPC framework is investigated. Consider the following multi-objective optimization problem:

$$\min_{U} J(U, \xi) \tag{5.13a}$$

$$\begin{aligned}&\text{subject to}\\ &\xi_{k+1} = h(\xi_k, \omega_k),\ \xi_0 = \xi,\\ &\xi_k \in \Xi,\ k = 1, 2, \ldots, N,\\ &\omega_k \in \Omega, k = 0, 1, \ldots, N-1,\end{aligned} \tag{5.13b}$$

with

$$\xi = \begin{bmatrix} \mathbf{x} \\ s \end{bmatrix}, \omega = \begin{bmatrix} \mathbf{u} \\ v_s \end{bmatrix}, h(\xi, \omega) = \begin{bmatrix} f_d(\mathbf{x}, \mathbf{u}) \\ g_d(s, v_s) \end{bmatrix}. \tag{5.14}$$

Here, (5.14) describes the discretized augmented system; $J(U, \xi) = [J_1(U, \xi), J_2(U, \xi)]^{\mathrm{T}}$ is a vector-valued objective function constructed for the two PF requirements; $U = \mathrm{col}(\omega_0, \omega_1, \ldots, \omega_{N-1})$ is the sequence of control inputs to be optimized; ξ_k denotes the k-step predicted state from initial condition $\xi_0 = \xi$; Ξ and Ω represent the constraints on state and input, respectively.

Each objective function is in the following form:

$$J_i(U, \xi) = \sum_{k=0}^{N-1} L_i(\xi_k, \omega_k) + E_i(\xi_N), \tag{5.15}$$

where

$$L_1(\xi, \omega) = \|\xi - \xi_p\|^2_{Q_1} + \|\omega\|^2_{R_1}, \tag{5.16a}$$

$$L_2(\xi, \omega) = \|\xi - \xi_t\|^2_{Q_2} + \|\omega\|^2_{R_2}, \tag{5.16b}$$

$$E_1(\xi) = \|\xi - \xi_p\|^2_{P_1}, \tag{5.16c}$$

$$E_2(\xi) = \|\xi - \xi_t\|^2_{P_2}. \tag{5.16d}$$

Here, $\xi_p = \mathrm{col}(p_{\mathbf{x}}(s), s)$ can be viewed as the reference for path convergence while $\xi_t = \mathrm{col}(p_{\mathbf{x}}(s_r), s_r)$ is the reference for speed assignment with s_r generated by integrating $\dot{s}_r$ from s_0; Q_i, R_i, and P_i are weighting matrices, positive definite.

Since $J_i(U, \xi)$ are in general conflicting with each other, we take advantage of the notion of Pareto optimality to measure the efficiency of a solution in the multi-objective optimization problem (Deb 2011):

Definition 5.1 (Pareto Optimality) Let $\bar{U}$ be a feasible point for (5.13). Then, $\bar{U}$ is said to be Pareto optimal (PO) if there is no other feasible point U such that $J_i(U, \xi) \leq J_i(\bar{U}, \xi)$ for all i, and $J_i(U, \xi) < J_i(\bar{U}, \xi)$ for at least one i, and $\bar{U}$ is said to be weakly PO if $J_i(U, \xi) < J_i(\bar{U}, \xi)$ for all i.

By definition, we shall notice that there usually exist more than one PO points. More specifically, all of the PO points constitute the so-called Pareto frontier. The complete solution of (5.13) is to determine the Pareto frontier using evolutionary methods (Deb 2011) or classical methods (Miettinen 2012). However, for the PF problem, it is unnecessary to calculate the entire frontier. What we need is just to find one preferred point on the Pareto frontier. Even a weakly PO point may be acceptable. In this regard, two methods which are capable of handling objective prioritization are investigated to solve the MOMPC PF problem.

5.3 Solve the MOMPC Problem

5.3.1 Weighted-Sum Method

The WS method (Miettinen 2012) scalarizes the vector-valued objective function by assigning a weight to each objective. Instead of solving (5.13) directly, the WS method solves the following single objective problem:

$$\begin{aligned} &\min_{U} \ J_W = a^{\mathrm{T}} J(U, \xi) \\ &\text{s.t.} \ \ (5.13\text{b}), \end{aligned} \tag{5.17}$$

where $a = [\alpha, 1 - \alpha]^{\mathrm{T}}$ with $0 \leq \alpha \leq 1$.

Remark 5.1 There are some comments on the WS method. First of all, the solution U^* of (5.17) is always weakly PO, or PO if $0 < \alpha < 1$ for (5.13) without any further assumptions (Miettinen 2012). This provides a solid theoretical support for use of the WS method. Second, by scalarizing the original problem, we can choose from a wide class of optimization algorithms, namely the gradient descent methods (Boyd et al. 2004), to efficiently solve the optimization problem. Recent developments of fast MPC implementation algorithms are exclusively designed for single objective optimizations (Diehl et al. 2009). Hence, the use of WS method serves as a link between the MOMPC and a vast majority of results for single objective MPC. Finally, through the use of WS method we can incorporate some priori knowledge into the MOMPC formulation by selecting preferred weights for each objective.

To automatically choose the appropriate weight for each objective function, we propose the following logistic function:

$$\alpha(Er) = \frac{1}{1 + e^{-\beta Er}}, \tag{5.18}$$

where $Er = \|\boldsymbol{\eta} - \boldsymbol{\eta}_p\|_K^2$ with $\boldsymbol{\eta}_p = [p_x(s), p_y(s), p_\psi(s)]^{\mathrm{T}}$ serves as an indicator of path convergence; $\beta > 0$ controls the change rate of the function; K is a positive definite weighting matrix. The logistic function (5.18) is smooth and monotonic. The range of (5.18) is $[0.5, 1)$ corresponding to the domain $[0, +\infty)$. Since the path convergence is more important than the speed assignment, the logistic function (5.18) always puts more weights on J_1.

We have selected the path dynamics to be a single integrator (5.12), therefore, the forward motion requirement can be formulated as constraints on the input of the augmented system. Further considering AUV thrust limits, we have the input constraint set Ω polyhedral:

$$\Omega = \{\omega \in \mathbb{R}^4 | \, \underline{\omega}(j) \leq \omega(j) \leq \bar{\omega}(j), \; j = 1, \ldots, 4\}, \tag{5.19}$$

where $\underline{\omega}(j)$ and $\bar{\omega}(j)$ represent the lower and upper bounds on the jth component, respectively. Suppose $s_0 = 0$ and $s_1 = +\infty$, and there is no constraint on the system state, i.e., $\Xi = \mathbb{R}^4$. The scalarized problem (5.17) can be solved by applying PMP with the help of barrier functions.

Define the Hamiltonian by

$$H(\xi, \lambda, \omega) = \ell(\xi, \omega) + \lambda^{\mathrm{T}} h(\xi, \omega) - \gamma(b_1(\omega) + b_2(\omega)), \tag{5.20}$$

with

$$\ell(\xi, \omega) = \alpha L_1(\xi, \omega) + (1 - \alpha) L_2(\xi, \omega), \tag{5.21}$$

$$b_1(\omega) = \sum_{j=1}^{4} \log(\omega(j) - \underline{\omega}(j)), \tag{5.22}$$

$$b_2(\omega) = \sum_{j=1}^{4} \log(\bar{\omega}(j) - \omega(j)). \tag{5.23}$$

Here, λ is called the costate, and $b_1(\omega)$, $b_2(\omega)$ are barriers for the inequality (5.19); α is the weight determined by (5.18) and γ is a small positive number. Define $E(\xi) = \alpha E_1(\xi) + (1 - \alpha) E_2(\xi)$, then PMP claims that for a local optimal control $\{\omega_i^*\}_{i=0}^{N-1}$, there exist $\{\lambda_i^*\}_{i=0}^{N}$ satisfying the following optimality conditions:

$$\xi_{k+1}^* = h(\xi_k^*, \omega_k^*), \tag{5.24a}$$

$$\lambda_i^* = \lambda_{i+1}^* + H_\xi^{\mathrm{T}}(\xi_i^*, \lambda_{i+1}^*, \omega_i^*)\Delta t, \tag{5.24b}$$

$$\lambda_N^* = E_\xi^{\mathrm{T}}(\xi_N^*), \tag{5.24c}$$

$$\xi_0^* = \xi, \tag{5.24d}$$

$$H_\omega(\xi_i^*, \lambda_{i+1}^*, \omega_i^*) = 0, \tag{5.24e}$$

which sufficiently solves the KKT conditions by imposing N boundary conditions (5.24b)–(5.24c). Observe that given the initial state ξ and a control input $U = \mathrm{col}(\omega_0, \ldots, \omega_{N-1})$, the states $\{\xi_i^*\}_{i=0}^{N}$ and costates $\{\lambda_i^*\}_{i=0}^{N}$ can be determined via recurrence relations (5.24a)–(5.24b). Therefore, at each sampling instant, we only need to solve equations (5.24e), which avoids expensive numerical operations in terms of successive linearizations in solving the KKT conditions (Boyd et al. 2004). Furthermore, fast MPC implementation algorithms that are based on solving the PMP system (5.24) such as C/GMRES Ohtsuka (2004); Shen et al. (2017) and Gradient Projection Graichen (2012) can, in principle, be used.

The proposed WS-based multi-objective model predictive control (WS-MOMPC) algorithm is summarized in Algorithm 5.1.

Algorithm 5.1 WS-MOMPC PF Algorithm

1: Input the objective functions $J_i(U, \xi)$ in (5.15).
2: Measure current state $\xi(t)$.
3: Evaluate the value of (5.18) using $Er(t)$.
4: Solve the scalarized problem (5.17) with $\xi = \xi(t)$.
5: Let $U^* = \text{col}(\omega_0^*, \dots, \omega_{N-1}^*)$ denote the solution.
6: Implement ω_0^* to the augmented system for one sampling period.
7: At next sampling time instant, set $t = t + \Delta t$, then repeat from step 2.

5.3.2 *Lexicographic Ordering Method*

The WS method is an indirect method which presumes that the value of each objective function at the solution is negatively correlated to its weight. Solving the nonlinear problem (5.17), however, only obtains a local minimum, which makes the negative correlation not strictly monotonic. Therefore, a direct method called LO is also studied for the MOMPC PF problem.

To solve the MOMPC problem (5.13), the LO method creates a procedure which considers the two objectives, one at a time, ordered by their priority Miettinen (2012); He et al. (2015). At each sampling time instant, the following optimization problems are solved sequentially:

$$J_1^*(\xi) = \min_U \{J_1(U, \xi) \mid (5.13\text{b})\}, \tag{5.25a}$$

$$J_2^*(\xi) = \min_U \{J_2(U, \xi) \mid (5.13\text{b}),\ J_1(U, \xi) = J_1^*(\xi)\}. \tag{5.25b}$$

Then we obtain the solution

$$U^* = \arg\min_U \{J_2(U, \xi) \mid (5.13\text{b}),\ J_1(U, \xi) = J_1^*(\xi)\}. \tag{5.26}$$

To prevent the numerical algorithm from stalling and to improve the computational efficiency, the equality constraint $J_1(U, \xi) = J_1^*(\xi)$ in (5.25b) is often replaced by the following inequality constraint:

$$J_1(U, \xi) \leq J_1^*(\xi) + \epsilon, \tag{5.27}$$

where $\epsilon \geq 0$ is the tolerance.

Remark 5.2 There are some comments on the LO method. Firstly, the solution (5.26) of the standard LO procedure is always PO (Miettinen 2012). When the relaxation (5.27) is included, the solution U^* becomes weakly PO. However, the weak Pareto optimality is practically acceptable for the AUV PF control problem. Also, we will see shortly in the next section that the relaxation is totally compatible with the path convergence conditions. Secondly, due to the imposed additional constraint (5.27) in the second layer optimization (5.25b), the PMP conditions can hardly be applied. Instead, we resort to direct methods which solve the KKT conditions.

The LO method and WS method are closely related. To explore the internal relationship, we simplify the mathematical expression of (5.17) as follows:

$$\begin{aligned} &\min_U \ J_W(U) = \alpha J_1(U) + (1-\alpha) J_2(U) \\ &\text{s.t.} \ \ c_i(U) = 0, \ \ m_j(U) \le 0. \end{aligned} \tag{5.28}$$

Also simplify the mathematical expression of (5.25b) with the relaxation (5.27) as

$$\begin{aligned} &\min_U \ J_2(U) \\ &\text{s.t.} \ \ c_i(U) = 0, \ \ m_j(U) \le 0, \ \ J_1(U) \le \delta, \end{aligned} \tag{5.29}$$

where $c_i(U)$ are the equality constraints including system dynamics and boundary conditions; $m_j(U)$ represent the inequality constraints including state constraints and control constraints; and $\delta = J_1^*(\xi) + \epsilon$.

Then we show the relationship between the two methods by explicating the KKT conditions.

Theorem 5.1 *If the weights for the WS method are selected precisely the same as the optimal values of dual variables for the LO method, the two methods yield the same optimal solution.*

Proof Let $\bar{J}_W(U) = \bar{\alpha} J_1(U) + J_2(U)$ with $\bar{\alpha} = \frac{\alpha}{1-\alpha}$. Obviously, minimizing J_W is equivalent to minimizing $\bar{J}_W$. Define the Lagrangian for (5.28) as

$$\bar{L}_W = J_2(U) + \Sigma \bar{\lambda}_i c_i(U) + \Sigma \bar{\mu}_j m_j(U) + \bar{\alpha} J_1(U),$$

where $\bar{\lambda}_i$ and $\bar{\mu}_j$ are dual variables.

Define the Lagrangian for (5.29) as

$$L_\delta = J_2(U) + \Sigma \lambda_i c_i(U) + \Sigma \mu_j m_j(U) + \nu (J_1(U) - \delta),$$

where λ_i, μ_j, and ν are dual variables. Then we list the detailed KKT conditions for (5.28) as follows:

$$\nabla_U \bar{L}_W(U^*) = 0, \ \ c_i(U^*) = 0, \ \ m_j(U^*) \le 0, \tag{5.30a}$$

$$\bar{\mu}_j^* m_j(U^*) = 0, \quad \bar{\mu}_j^* \geq 0, \tag{5.30b}$$

where ∇_U represents the gradient with respect to U. Let us also list the detailed KKT conditions for problem (5.29)

$$\nabla_U L_\delta(U^*) = 0, \quad c_i(U^*) = 0, \quad m_j(U^*) \leq 0, \tag{5.31a}$$

$$\mu_j^* m_j(U^*) = 0, \quad \mu_j^* \geq 0, \tag{5.31b}$$

$$J_1(U^*) - \delta \leq 0, \tag{5.31c}$$

$$\nu^*(J_1(U^*) - \delta) = 0, \quad \nu^* \geq 0. \tag{5.31d}$$

Further expanding $\nabla_U \bar{L}_W(U^*)$ and $\nabla_U L_\delta(U^*)$, we have

$$\nabla J_2(U^*) + \Sigma \bar{\lambda}_i^* \nabla c_i(U^*) + \Sigma \bar{\mu}_j^* \nabla m_j(U^*) + \bar{\alpha} \nabla J_1(U^*) = 0,$$

$$\nabla J_2(U^*) + \Sigma \lambda_i^* \nabla c_i(U^*) + \Sigma \mu_j^* \nabla m_j(U^*) + \nu^* \nabla J_1(U^*) = 0.$$

We notice that the main differences between the above two sets of KKT conditions are the presence of (5.31c) and (5.31d). If $\bar{\alpha} = \nu^*$, (5.30) and (5.31) yield the same optimal solution U^* with $\bar{\lambda}_i^* = \lambda_i^*$ and $\bar{\mu}_j^* = \mu_j^*$ due to the fact that:
Case 1: The last inequality is active, i.e., $J_1(U^*) = \delta$. Then (5.31c) and (5.31d) become solely $\nu^* \geq 0$. By definition we have $\bar{\alpha} \geq 0$, therefore, in this case the two sets of KKT conditions are exactly the same.
Case 2: The last inequality is inactive, i.e., $J_1(U^*) < \delta$. Then (5.31c) and (5.31d) become $\nu^* = 0$. So we have $\bar{\alpha} = \nu^* = 0$, which make the two sets of KKT conditions identical.

Remark 5.3 As seen from the above proof, the WS method and LO method are implicitly related by the optimal value of dual variables. Therefore, it is difficult to find the corresponding ϵ which generates the same solution as that using WS method with the change of α obeying (5.18).

The proposed LO-based multi-objective model predictive control (LO-MOMPC) algorithm is summarized in Algorithm 5.2.

Algorithm 5.2 LO-MOMPC PF Algorithm

1: Input the objective functions $J_i(U, \xi)$ in (5.15).
2: Measure current state $\xi(t)$.
3: Sequentially solve the LO subproblems (5.25) with $\xi = \xi(t)$.
4: Let $U^* = \text{col}(\omega_0^*, \ldots, \omega_{N-1}^*)$ be the solution of the second layer problem (5.25b).
5: Implement ω_0^* to the augmented system for one sampling period.
6: At next sampling time instant, set $t = t + \Delta t$, then repeat from step 2.

5.4 Convergence Analysis

In order to follow the standard analysis procedure used in conventional MPC Rawlings et al. (2017), the original MOMPC problem (5.13) needs to be modified and reformulated as the regulation problem of the error dynamics:

$$\min_{\tilde{U}} J(\tilde{U}, \tilde{\xi}) \tag{5.32a}$$

$$\begin{aligned} &\text{subject to} \\ &\tilde{\xi}_{k+1} = \tilde{h}(\tilde{\xi}_k, \tilde{\omega}_k),\ \tilde{\xi}_0 = \tilde{\xi}, \\ &\tilde{\xi}_k \in \tilde{\Xi},\ k = 1, 2, \ldots, N, \\ &\tilde{\omega}_k \in \tilde{\Omega}, k = 0, 1, \ldots, N-1, \end{aligned} \tag{5.32b}$$

where $\tilde{h}$ denotes the error dynamics; $J(\tilde{U}, \tilde{\xi}) = [J_1, J_2]^{\mathrm{T}}$ with $J_i = \sum_{k=0}^{N-1} L_i$ $(\tilde{\xi}_k, \tilde{\omega}_k) + E_i(\tilde{\xi}_N)$; $\tilde{\Xi}$ and $\tilde{\Omega}$ denote system constraints.

Then the convergence of the MOMPC solution can be analyzed with the following assumptions:

Assumption 5.1 The functions $\tilde{h}(\cdot)$, $L_i(\cdot)$ and $E_i(\cdot)$ are continuous; $\tilde{h}(0, 0) = 0$, $L_i(0, 0) = 0$ and $E_i(0) = 0$.

Assumption 5.2 The sets $\tilde{\Xi}$ and $\tilde{\Pi}$ are closed, and $\tilde{\Omega}$ is compact. Each set contains the origin in its interior.

Assumption 5.3 There exist an invariant set $\tilde{\Pi}$ for the error dynamics $\tilde{h}$ and a local controller $\kappa(\tilde{\xi})$ such that

$$\begin{aligned} &\kappa(\tilde{\xi}) \in \tilde{\Omega}, \quad \tilde{h}(\tilde{\xi}, \kappa(\tilde{\xi})) \in \tilde{\Pi}, \quad i = 1, 2, \\ &E_i(\tilde{h}(\tilde{\xi}, \kappa(\tilde{\xi}))) - E_i(\tilde{\xi}) + L_i(\tilde{\xi}, \kappa(\tilde{\xi})) \leq 0, \end{aligned} \tag{5.33}$$

for any $\tilde{\xi} \in \tilde{\Pi}$.

For the WS-MOMPC scheme, we need to modify the problem by imposing the terminal constraints as follows:

$$J_W^*(\tilde{\xi}, \alpha) = \min_{\tilde{U}} \{a^{\mathrm{T}} J(\tilde{U}, \tilde{\xi}) \mid (5.32\text{b}), \tilde{\xi}_N \in \tilde{\Pi}\}. \tag{5.34}$$

Also, before solving (5.34), at each sampling instant (except for the first time $t = 0$), a subproblem that determines the appropriate weight α has to be solved:

$$\alpha^*(\tilde{\xi}, \alpha_d, J_\alpha) = \arg\min_{\alpha} f_\alpha(\alpha - \alpha_d) \tag{5.35a}$$

$$\begin{aligned} \text{s.t. } &J_W^*(\tilde{\xi}, \alpha) \leq J_\alpha, \\ &0 \leq \alpha \leq 1, \end{aligned} \tag{5.35b}$$

where α_d is the target weight calculated by (5.18); f_a is a convex function that measures the distance between α and α_d; Let $\tilde{U}^*_{-1} = \text{col}(\tilde{\omega}^*_{-1,0}, ..., \tilde{\omega}^*_{-1,N-1})$ and a^*_{-1} be the previous time solutions for (5.34) and (5.35), respectively. Then $J_\alpha = a^{*\text{T}}_{-1} J(\tilde{U}^0, \tilde{\xi})$ with $\tilde{U}^0 = \text{col}(\tilde{\omega}^*_{-1,1}, ..., \tilde{\omega}^*_{-1,N-1}, \kappa(\tilde{\xi}_{-1,N}))$.

Then we can have the following results on the convergence of the solution:

Theorem 5.2 *Suppose Assumptions 5.1–5.3 are satisfied, and the problem (5.34) is feasible at time $t = 0$. Then with the WS-MOMPC algorithm, the error state converges to the origin, i.e., $\tilde{\xi} \to 0$ as $t \to \infty$.*

Proof The recursive feasibility can be ensured because the shifted control sequence $\tilde{U}^0$ is always feasible for (5.34) and $a^{*\text{T}}_{-1}$ is always feasible for (5.35) at the next time instant.

Evaluating the optimal value function of J^*_W at two successive time instants, we have

$$\begin{aligned}
\Delta J_W &= J_W(\tilde{U}^*, \tilde{\xi}) - J_W(\tilde{U}^*_{-1}, \tilde{\xi}_{-1,0}) \\
&= a^{*\text{T}} J(\tilde{U}^*, \tilde{\xi}) - a^{*\text{T}}_{-1} J(\tilde{U}^*_{-1}, \tilde{\xi}_{-1,0}) \\
&\leq a^{*\text{T}}_{-1} J(\tilde{U}^0, \tilde{\xi}) - a^{*\text{T}}_{-1} J(\tilde{U}^*_{-1}, \tilde{\xi}_{-1,0}) \\
&= \alpha^*_{-1}[E_1(\tilde{h}(\tilde{\xi}_{-1,N}, \kappa(\tilde{\xi}_{-1,N}))) - E_1(\tilde{\xi}_{-1,N}) + L_1(\tilde{\xi}_{-1,N}, \kappa(\tilde{\xi}_{-1,N})) - L_1(\tilde{\xi}_{-1,0}, \tilde{\omega}^*_{-1,0})] \\
&\quad +(1 - \alpha^*_{-1})[E_2(\tilde{h}(\tilde{\xi}_{-1,N}, \kappa(\tilde{\xi}_{-1,N}))) - E_2(\tilde{\xi}_{-1,N}) + L_2(\tilde{\xi}_{-1,N}, \kappa(\tilde{\xi}_{-1,N})) - L_2(\tilde{\xi}_{-1,0}, \tilde{\omega}^*_{-1,0})].
\end{aligned}$$

By Assumption 5.3, $\Delta J_W \leq -\alpha^*_{-1} L_1(\tilde{\xi}_{-1,0}, \tilde{\omega}^*_{-1,0}) - (1 - \alpha^*_{-1}) L_2(\tilde{\xi}_{-1,0}, \tilde{\omega}^*_{-1,0}) \leq 0$. By construction, we have $J_W \geq 0$. Therefore, $J^*_W(t)$ is a nonincreasing sequence and lower bounded by zero. By contradiction, we have $-\alpha^*_{-1} L_1(\tilde{\xi}_{-1,0}, \tilde{\omega}^*_{-1,0}) - (1 - \alpha^*_{-1}) L_2(\tilde{\xi}_{-1,0}, \tilde{\omega}^*_{-1,0}) \to 0$ as $t \to \infty$. Since $Q_i > 0$ and $R_i > 0$ the convergence of the error state can be guaranteed, i.e., $\tilde{\xi} \to 0$ as $t \to \infty$.

For the LO-MOMPC PF scheme, the convergence of the MOMPC solution can be guaranteed with the following assumption:

Assumption 5.4 There exist an invariant set $\tilde{\Pi}$ for the error dynamics $\tilde{h}$, containing the origin in the interior, and a local feedback control $\kappa_1(\tilde{\xi})$ such that

$$\begin{aligned}
&\kappa_1(\tilde{\xi}) \in \tilde{\Omega}, \quad \tilde{h}(\tilde{\xi}, \kappa_1(\tilde{\xi})) \in \tilde{\Pi}, \\
&E_1(\tilde{h}(\tilde{\xi}, \kappa_1(\tilde{\xi}))) - E_1(\tilde{\xi}) + L_1(\tilde{\xi}, \kappa_1(\tilde{\xi})) \leq 0,
\end{aligned} \tag{5.36}$$

for any $\tilde{\xi} \in \tilde{\Pi}$.

We modify the LO-MOMPC problem by imposing the terminal constraints:

$$J_1^*(\tilde{\xi}) = \min_{\tilde{U}} \{J_1(\tilde{U}, \tilde{\xi}) \mid (5.32\text{b}), \ \tilde{\xi}_N \in \tilde{\Pi}\}, \tag{5.37a}$$

$$J_2^*(\tilde{\xi}) = \min_{\tilde{U}} \left\{ J_2(\tilde{U}, \tilde{\xi}) \mid J_1(\tilde{U}, \tilde{\xi}) \leq J_1^*(\tilde{\xi}) + \epsilon, \ (5.32\text{b}), \ \tilde{\xi}_N \in \tilde{\Pi} \right\}. \tag{5.37b}$$

Then we have the following results on the convergence of the solution:

Theorem 5.3 *Suppose Assumptions 5.1, 5.2, and 5.4 are satisfied, and for* $\epsilon = 0$ *the problem (5.37) is feasible at time* $t = 0$. *Let* $\tilde{\xi}_{-1,0}$ *denote the previous system state and* $\tilde{\omega}^*_{-1,0}$ *denote the first element of the previous time solution. For* $t > 0$, *if we choose* $\epsilon = L_1(\tilde{\xi}_{-1,0}, \tilde{\omega}^*_{-1,0}) - \tilde{\xi}^{\mathrm{T}} Q_\epsilon \tilde{\xi}$ *with* $Q_1 > Q_\epsilon > 0$, *then the error state converges to the origin, i.e.,* $\tilde{\xi} \to 0$ *as* $t \to \infty$.

Proof Let $\tilde{U}^*_{-1} = \mathrm{col}(\tilde{\omega}^*_{-1,0}, \ldots, \tilde{\omega}^*_{-1,N-1})$ denote the solution for the previous sampling instant and $\tilde{\xi}_{-1,i}$ denote the corresponding state prediction for $i = 0, 1, \ldots, N$. An initial guess $\tilde{U}^0 = \mathrm{col}(\tilde{\omega}^*_{-1,1}, \ldots, \tilde{\omega}^*_{-1,N-1}, \kappa_1(\tilde{\xi}_{-1,N}))$ can be constructed for the first layer subproblem (5.37a), which admits the horizontal feasibility. The hierarchical feasibility is also preserved since the solution of (5.37a) is always feasible for (5.37b).

Evaluating the optimal value function of J_1^* at these two successive sampling instants, we have

$$
\begin{aligned}
\Delta J_1 &= J_1(\tilde{U}^*, \tilde{\xi}) - J_1(\tilde{U}^*_{-1}, \tilde{\xi}_{-1,0}) \\
&\leq J_1(\tilde{U}^0, \tilde{\xi}) - J_1(\tilde{U}^*_{-1}, \tilde{\xi}_{-1,0}) + \epsilon \\
&= E_1(\tilde{h}(\tilde{\xi}_{-1,N}, \kappa_1(\tilde{\xi}_{-1,N}))) - E_1(\tilde{\xi}_{-1,N}) + L_1(\tilde{\xi}_{-1,N}, \kappa_1(\tilde{\xi}_{-1,N})) - \tilde{\xi}^{\mathrm{T}} Q_\epsilon \tilde{\xi}.
\end{aligned}
\tag{5.38}
$$

With Assumption 5.4 we have $\Delta J_1 \leq -\tilde{\xi}^{\mathrm{T}} Q_\epsilon \tilde{\xi} \leq 0$. By construction, we know that $J_1 \geq 0$. Therefore, $J_1^*(t)$ is a nonincreasing sequence and lower bounded by zero. By contradiction, we have $-\tilde{\xi}^{\mathrm{T}} Q_\epsilon \tilde{\xi} \to 0$ as $t \to \infty$. Since $Q_\epsilon > 0$, the convergence of the error state trajectory to the origin can be guaranteed, i.e., $\tilde{\xi} \to 0$ as $t \to \infty$.

Remark 5.4 The differences between Assumptions 5.3 and 5.4 reflect the underlying differences between the WS-MOMPC and LO-MOMPC schemes. Assumption 5.3 requires that for each and every objective J_i we need to find the same local feedback control law $\kappa(\tilde{\xi})$ and terminal region $\tilde{\Pi}$ such that the inequalities in (5.33) can hold. This will guarantee both path convergence and speed assignment but complicate the MPC design at the same time. In terms of the path-following control problem, one possible reformulation that satisfies Assumption 5.3 is the stringent fulfillment of the speed assignment along the ZPE manifold. In this way, however, the PF problem is solved as a trajectory-tracking problem with two performance indices that emphasize different error states (position vs. forward speed), and thus the extra degree of freedom of adjusting $s(t)$ is lost. In contrast, Assumption 5.4 requires to find the local feedback control law and terminal region for the most important objective J_1 such that the inequality in (5.36) can hold. This will only rigorously guarantee the path convergence which is the primary task in the PF problem. However, since the path convergence and speed assignment are physically achievable and the relaxation parameter ϵ is incorporated, the control on speed assignment is often satisfactory. Adopting LO-MOMPC scheme simplifies the MPC design and retains the flexibility of adjusting $s(t)$ on the fly.

Remark 5.5 There are different ways to reformulate the problem as the regulation of the error dynamics and to satisfy the assumptions. As an example, we can take advantage of the ZPE manifold. Since the ZPE manifold (5.11) serves as a reference for the AUV to track, the following relation holds:

$$\dot{p}_x = p_u \cos p_\psi - p_v \sin p_\psi, \tag{5.39a}$$
$$\dot{p}_y = p_u \sin p_\psi + p_v \cos p_\psi, \tag{5.39b}$$
$$\dot{p}_\psi = p_r. \tag{5.39c}$$

Decomposing the kinematic error $\tilde{\boldsymbol{\eta}}$ in the vessel parallel reference frame, we have

$$\tilde{\boldsymbol{\eta}} = \begin{bmatrix} \tilde{x} \\ \tilde{y} \\ \tilde{\psi} \end{bmatrix} = \begin{bmatrix} -\cos\psi & -\sin\psi & 0 \\ \sin\psi & -\cos\psi & 0 \\ 0 & 0 & -1 \end{bmatrix} \begin{bmatrix} x - p_x \\ y - p_y \\ \psi - p_\psi \end{bmatrix}. \tag{5.40}$$

Define the velocity error $\tilde{\mathbf{v}}$ in the following way:

$$\tilde{\mathbf{v}} = \begin{bmatrix} \tilde{u} \\ \tilde{v} \\ \tilde{r} \end{bmatrix} = \begin{bmatrix} u - p_u \cos\tilde{\psi} + p_v \sin\tilde{\psi} \\ v - p_u \sin\tilde{\psi} - p_v \cos\tilde{\psi} \\ r - p_r \end{bmatrix}. \tag{5.41}$$

Further define $\tilde{\boldsymbol{\tau}} = \text{col}(\tilde{\tau}_1, \tilde{\tau}_2, \tilde{\tau}_3)$ in which

$$\begin{aligned}\tilde{\tau}_1 = {} & \tfrac{F_u}{M_{\dot{u}}} + \tfrac{M_{\dot{v}}}{M_{\dot{u}}}(\tilde{v} + p_u \sin\tilde{\psi} + p_v \cos\tilde{\psi})(\tilde{r} + p_r) - \tfrac{X_u}{M_{\dot{u}}}(\tilde{u} + p_u \cos\tilde{\psi} - p_v \sin\tilde{\psi}) \\ & - \tfrac{D_u}{M_{\dot{u}}}(\tilde{u} + p_u \cos\tilde{\psi} - p_v \sin\tilde{\psi})|\tilde{u} + p_u \cos\tilde{\psi} - p_v \sin\tilde{\psi}| - (\dot{p}_u \cos\tilde{\psi} \\ & + p_u \sin\tilde{\psi}\tilde{r} - \dot{p}_v \sin\tilde{\psi} + p_v \cos\tilde{\psi}\tilde{r}),\end{aligned} \tag{5.42}$$

$$\begin{aligned}\tilde{\tau}_2 = {} & \tfrac{F_v}{M_{\dot{v}}} - \tfrac{M_{\dot{u}}}{M_{\dot{v}}}(\tilde{u} + p_u \cos\tilde{\psi} - p_v \sin\tilde{\psi})(\tilde{r} + p_r) - \tfrac{Y_v}{M_{\dot{v}}}(\tilde{v} + p_u \sin\tilde{\psi} + p_v \cos\tilde{\psi}) \\ & - \tfrac{D_v}{M_{\dot{v}}}(\tilde{v} + p_u \sin\tilde{\psi} + p_v \cos\tilde{\psi})|\tilde{v} + p_u \sin\tilde{\psi} + p_v \cos\tilde{\psi}| - (\dot{p}_u \sin\tilde{\psi} \\ & - p_u \cos\tilde{\psi}\tilde{r} + \dot{p}_v \cos\tilde{\psi} + p_v \sin\tilde{\psi}\tilde{r}),\end{aligned} \tag{5.43}$$

$$\begin{aligned}\tilde{\tau}_3 = {} & \tfrac{F_r}{M_{\dot{r}}} - \tfrac{M_{\dot{v}} - M_{\dot{u}}}{M_{\dot{r}}}(\tilde{u} + p_u \cos\tilde{\psi} - p_v \sin\tilde{\psi})(\tilde{v} + p_u \sin\tilde{\psi} \\ & + p_v \cos\tilde{\psi}) - \tfrac{N_r}{M_{\dot{r}}}(\tilde{r} + p_r) - \tfrac{D_r}{M_{\dot{r}}}(\tilde{r} + p_r)|\tilde{r} + p_r| - \dot{p}_r.\end{aligned} \tag{5.44}$$

Then it can be shown that

$$\dot{\tilde{\mathbf{x}}} = \begin{bmatrix} \dot{\tilde{x}} \\ \dot{\tilde{y}} \\ \dot{\tilde{\psi}} \\ \dot{\tilde{u}} \\ \dot{\tilde{v}} \\ \dot{\tilde{r}} \end{bmatrix} = \begin{bmatrix} p_r\tilde{y} - \tilde{u} + \tilde{y}\tilde{r} \\ -p_r\tilde{x} - \tilde{v} - \tilde{x}\tilde{r} \\ -\tilde{r} \\ \tilde{\tau}_1 \\ \tilde{\tau}_2 \\ \tilde{\tau}_3 \end{bmatrix} \triangleq \tilde{f}(\tilde{\mathbf{x}}, \tilde{\boldsymbol{\tau}}), \tag{5.45}$$

and at the origin $(\mathbf{0}, \mathbf{0})$, it admits that

$$\dot{p}_u = \frac{M_{\dot{v}}}{M_{\dot{u}}} p_v p_r - \frac{X_u}{M_{\dot{u}}} p_u - \frac{D_u}{M_{\dot{u}}} p_u |p_u| + \frac{F_u}{M_{\dot{u}}}, \tag{5.46a}$$

$$\dot{p}_v = -\frac{M_{\dot{u}}}{M_{\dot{v}}} p_u p_r - \frac{Y_v}{M_{\dot{v}}} p_v - \frac{D_v}{M_{\dot{v}}} p_v |p_v| + \frac{F_v}{M_{\dot{v}}}, \tag{5.46b}$$

$$\dot{p}_r = \frac{M_{\dot{u}} - M_{\dot{v}}}{M_{\dot{r}}} p_u p_v - \frac{N_r}{M_{\dot{r}}} p_r - \frac{D_r}{M_{\dot{r}}} p_r |p_r| + \frac{F_r}{M_{\dot{r}}}. \tag{5.46c}$$

Comparing (5.39) and (5.46) with (2.34) and (2.32), we find that the ZPE manifold can be viewed as the state trajectory generated by a virtual AUV owning exactly the same kinematic and dynamic properties as the real AUV. Since $(\mathbf{0}, \mathbf{0})$ indicates that the AUV is already on the manifold, the control input F_u, F_v and F_r will make the AUV stay on the manifold.

Detailed derivation is similar to (3.7)–(3.14).

Define $\tilde{\xi} = \mathrm{col}(\tilde{\boldsymbol{\eta}}, \tilde{\mathbf{v}}, 0)$, $\tilde{\omega} = \mathrm{col}(\tilde{\boldsymbol{\tau}}, 0)$ and the error dynamics $\tilde{h} = \mathrm{col}(\tilde{f}, g)$. We modify J_1 with

$$L_1(\tilde{\xi}, \tilde{\omega}) = \|\tilde{\xi}\|_{Q_1}^2 + \|\tilde{\omega}\|_{R_1}^2, \quad E_1(\tilde{\xi}) = \|\tilde{\xi}\|_{P_1}^2. \tag{5.47}$$

Leave L_2 and E_2 the same as in (5.16). In this way, Assumption 5.1 can be satisfied.

Notice that the error control signal construction (5.42)–(5.44) needs the values of $\dot{p}_u$, $\dot{p}_v$, and $\dot{p}_r$. By (5.11), we have $\dot{p}_v = 0$ and

$$\dot{p}_u = (p'_x p''_x + p'_y p''_y)(p'^2_x + p'^2_y)^{-\frac{1}{2}} \dot{s}^2 + (p'^2_x + p'^2_y)^{\frac{1}{2}} \ddot{s}, \tag{5.48a}$$

$$\begin{aligned} \dot{p}_r = & \frac{(p'^2_x + p'^2_y)(p'_x p'''_y - p'_y p'''_x) - 2(p'_x p''_y - p'_y p''_x)(p'_x p''_x + p'_y p''_y)}{(p'^2_x + p'^2_y)^2} \dot{s}^2 \\ & + \frac{p'_x p''_y - p'_y p''_x}{p'^2_x + p'^2_y} \ddot{s}. \end{aligned} \tag{5.48b}$$

Since the above calculations require the information of $\ddot{s}$, the path dynamics need to be modeled as a second order integrator, i.e.,

$$\dot{z} = \tilde{A} z + \tilde{B} v_s, \quad \tilde{A} = \begin{bmatrix} 0 & 1 \\ 0 & 0 \end{bmatrix}, \quad \tilde{B} = \begin{bmatrix} 0 \\ 1 \end{bmatrix}, \tag{5.49}$$

where $z = [s, \dot{s}]^{\mathrm{T}}$. Accordingly, the forward motion requirement will be formulated as $z_2 = \dot{s} \geq 0$. We further employ the manifold (5.11) as the terminal constraint, i.e., $\tilde{\Pi} = \{0\}$. Then Assumption 5.2 and Assumption 5.4 can be satisfied with $\kappa_1(0) = 0$.

5.5 Simulation Study

In this section, we present the simulation results of the Falcon AUV to follow a sinusoidal path $p_x(s) = s$ and $p_y(s) = \sin(s)$ with $s \geq 0$ in the local level plane. For the speed assignment, instead of directly setting a preferred path velocity $\dot{s}_r$, we set a desired surge velocity $u_r = 1$ (m/s) for the vehicle. This is a common selection in engineering practice because for AUVs with control surfaces (fins), the surge velocity effectively influences the actuation efficiency of the control surface (higher speed with higher reactivity). Then, according to (5.11b), we have the following explicit expression:

$$\dot{s}_r = u_r(p_x'^2 + p_y'^2)^{-\frac{1}{2}}, \tag{5.50}$$

to calculate ξ_t at each sampling instant.

All the simulations are based on the experimentally identified dynamic model of Falcon. The model parameter details can be found in Table 2.1. To solve the PMP equation (5.24e) the trust-region-dogleg algorithm (Conn et al. 2000) is adopted. To direct solve the NLP problem with KKT conditions, the embedded sequential quadratic programming algorithm (Boyd et al. 2004) of MATLAB function `fmincon` is used.

5.5.1 PF Control Using WS-MOMPC

For the WS-MOMPC PF control, we select the parameters as follows: The sampling period $\Delta t = 0.1$s, prediction horizon $N = 10$, $\gamma = 10^{-4}$, $K = I$, and the weighting matrices $Q_1 = \text{diag}(10^5, 10^5, 10^2, 0.1, 0.1, 0.1, 0.1])$, $Q_2 = \text{diag}(1, 1, 1, 10^3, 0.1, 0.1, 0.1)$, $P_1 = \text{diag}(100, 100, 10, 10^{-3}, 10^{-3}, 10^{-3}, 10^{-3})$, $P_2 = \text{diag}(0.1, 0.1, 0.1, 100, 10^{-3}, 10^{-3}, 10^{-3})$, $R_1 = R_2 = \text{diag}(10^{-3}, 10^{-3}, 10^{-3}, 10^{-3})$. The control limits $F_{u,\max} = 500$ (N), $F_{v,\max} = 500$ (N), and $F_{r,\max} = 500$ (Nm), and initial conditions $\mathbf{x}(0) = [0.5, 0, 0, 0, 0, 0]^{\mathrm{T}}$ and $s(0) = 0$.

The AUV path-following results with different β values are shown in Fig. 5.1. It can be observed that (i) for all of the three cases, the AUV trajectory successfully converges to the desired sinusoidal path, which validates the effectiveness of the proposed WS-MOMPC PF method; (ii) for each case, at the beginning, the AUV moves in the direction that is almost perpendicular to the path in order to get the fastest convergence, which is a desirable property; (iii) the larger β value results in the faster path convergence as the β value accounts for the sensitivity (slope) of the logistic function.

Figure 5.2 records the surge velocity of the AUV during the simulation (for the case of $\beta = 1$). As we can see, the surge velocity keeps very well at the desired speed after the sacrifice for path convergence in the beginning. In Fig. 5.3, the general thrust forces and moments as well as the imaginary path control inputs are plotted. As expected, they are all within the corresponding ranges of permitted values, which validates the effectiveness of barrier functions in the PMP-based implementation.

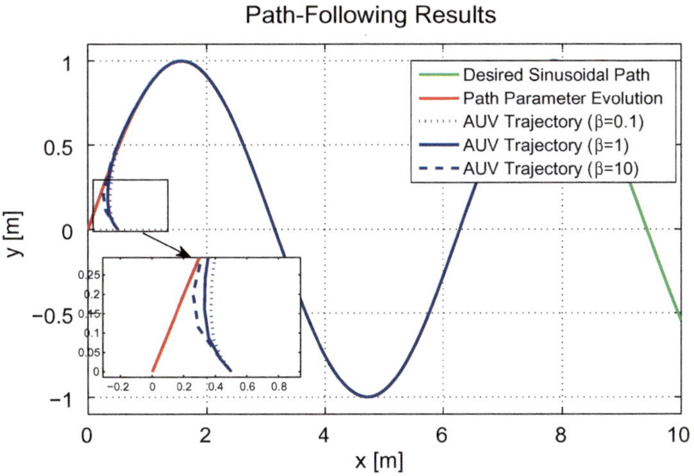

Fig. 5.1 AUV PF results with WS-MOMPC

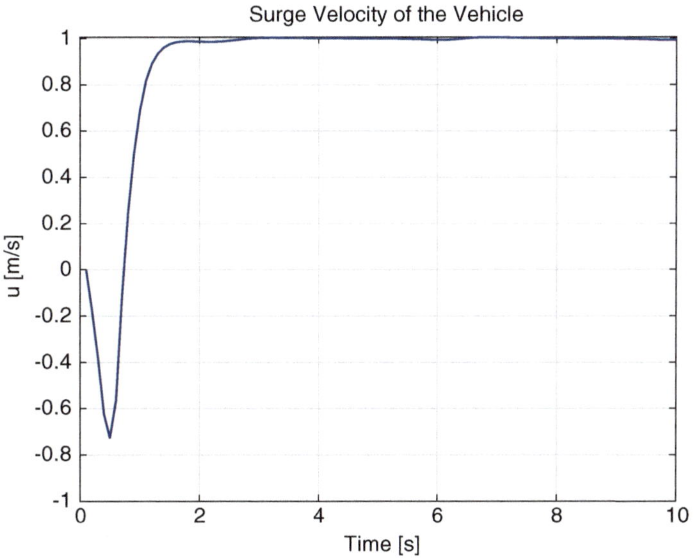

Fig. 5.2 Surge velocity of the AUV (WS-MOMPC)

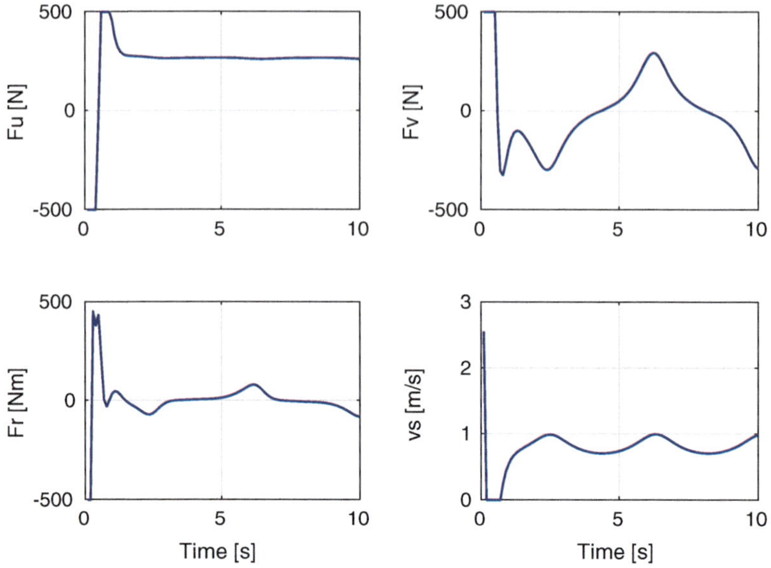

Fig. 5.3 Control inputs of the augmented system (WS-MOMPC)

5.5.2 PF Control Using LO-MOMPC

For the LO-MOMPC scheme, we simulate the AUV PF control using the way suggested in Remark 5.5. The parameters are chosen as follows: The sampling period $\Delta t = 0.1$s, prediction horizon $N = 10$, and weighting matrices $Q_1 = \text{diag}(10^5, 10^5, 10^3, 10^3, 10^{-3}, 10^{-3}, 10^{-3})$, $Q_2 = \text{diag}(10^4, 10^4, 10^2, 10^3, 10^{-3}, 10^{-3}, 10^3)$, $P_1 = \text{diag}(100, 100, 10, 100, 10^{-3}, 10^{-3}, 10^{-3})$, $P_2 = \text{diag}(100, 100, 10, 100, 10^{-3}, 10^{-3}, 10^2)$, $R_1 = R_2 = \text{diag}(10^{-3}, 10^{-3}, 10^{-3}, 10^{-3})$. The control limits $F_{u,\max} = 500$ (N), $F_{v,\max} = 500$ (N), and $F_{r,\max} = 500$ (Nm), and initial conditions $\mathbf{x}(0) = [0.5, 0, 0, 0, 0, 0]^{\mathrm{T}}$ and $s(0) = 0$.

The AUV path-following results are shown in Fig. 5.4. Similar observations can be made: (i) The AUV trajectory successfully converges to the desired sinusoidal path, which validates the effectiveness of the proposed LO-MOMPC PF method; (ii) in the beginning, the AUV moves in the direction almost perpendicular to the desired path in order to get the fastest convergence, which reflects the different priorities of the two PF requirements.

Figure 5.5 records the surge velocity of the AUV during the simulation. The surge velocity keeps very well at the desired speed after the initial sacrifice for path convergence. In Fig. 5.6, the general thrust forces and moments as well as the imaginary path control input are plotted.

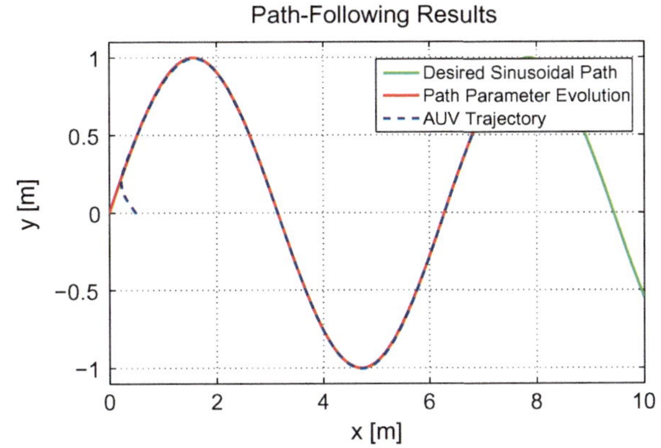

Fig. 5.4 AUV PF results using LO-MOMPC

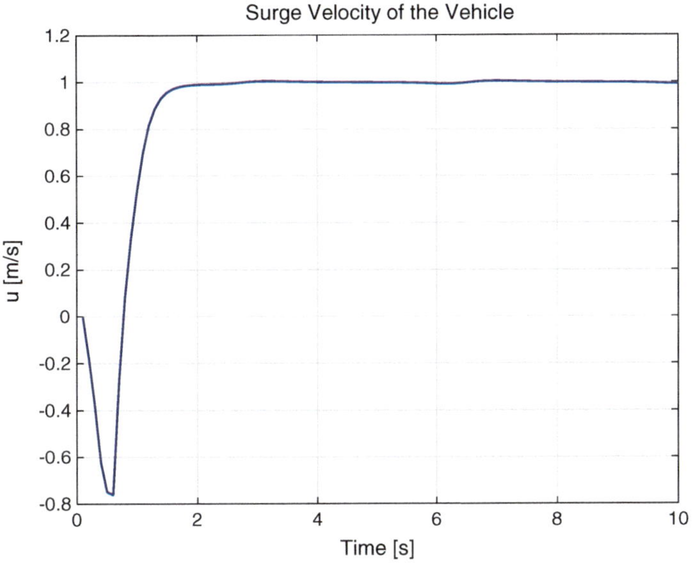

Fig. 5.5 Surge velocity of the AUV (LO-MOMPC)

Also the optimal values for the two objective functions are plotted in Fig. 5.7. It can be seen that J_1^* decreases monotonically, which indicates that the path convergence can be obtained; while J_2^* admits an initial hike, which conforms to the strict prioritization between the two control objectives.

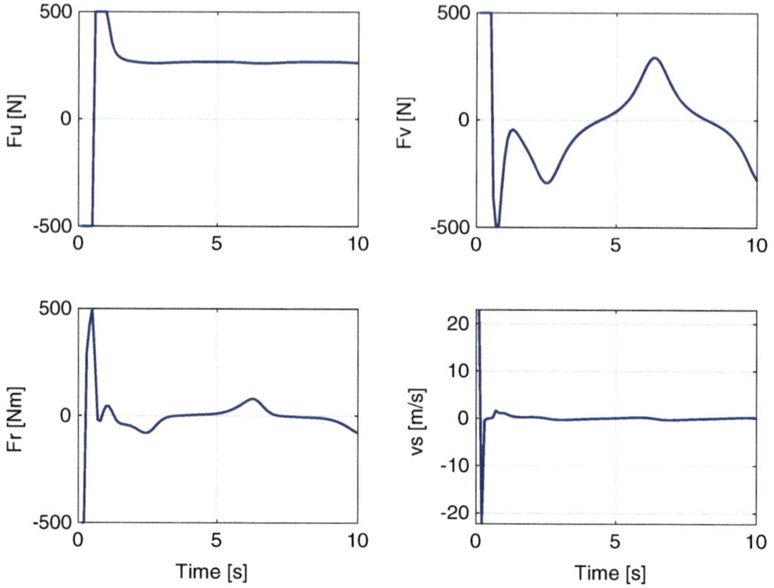

Fig. 5.6 Control inputs of the augmented system (LO-MOMPC)

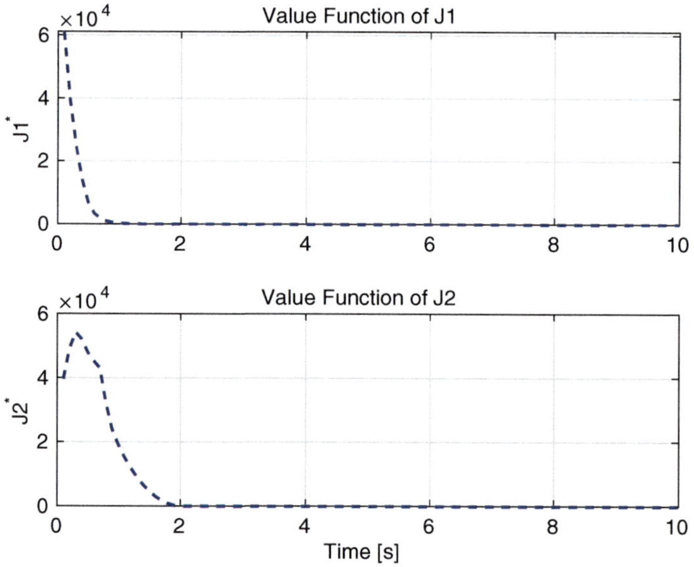

Fig. 5.7 Optimal value functions

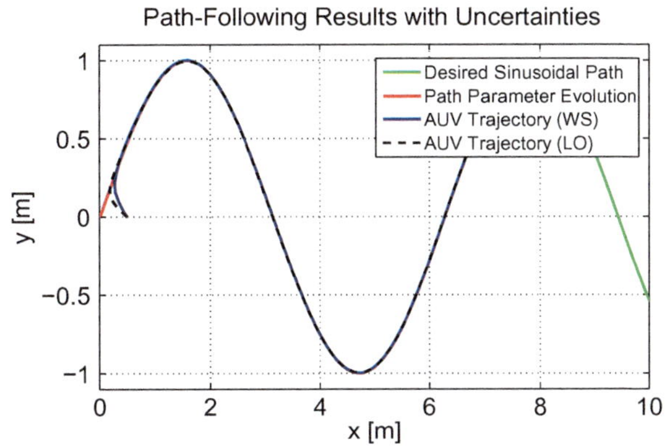

Fig. 5.8 AUV PF results (with parametric uncertainties)

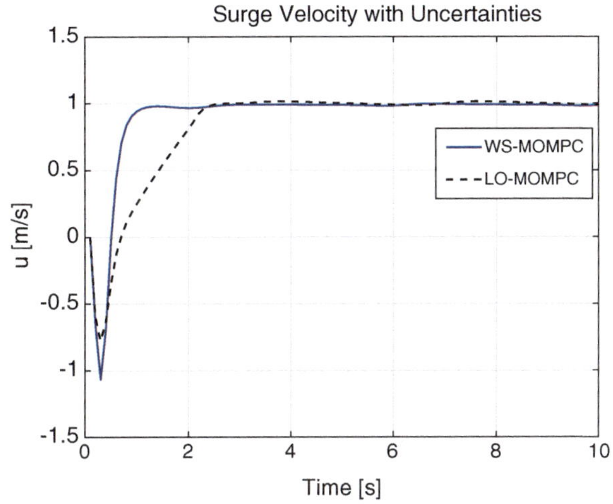

Fig. 5.9 Surge velocity of the AUV (with parametric uncertainties)

5.5.3 *Robustness Test*

The receding horizon control paradigm makes MPC robust to parametric uncertainty of the system model. This feature is extremely suitable for control of underwater vehicles whose dynamic model is always with uncertainties due to the poor knowledge of hydrodynamic coefficients. In Fig. 5.8, the path-following control is performed using a dynamic model with 20% model errors. As we can see, the AUV still closely follows the desired path, and the control performance is quite satisfactory. The surge velocity of the vehicle is plotted in Fig. 5.9, which well demonstrates the robustness of the proposed MOMPC method.

In addition, the robustness is also tested through the path-following task of non-differentiable paths since these types of path (e.g., the lawn mower pattern) represent actual oceanographic survey applications.

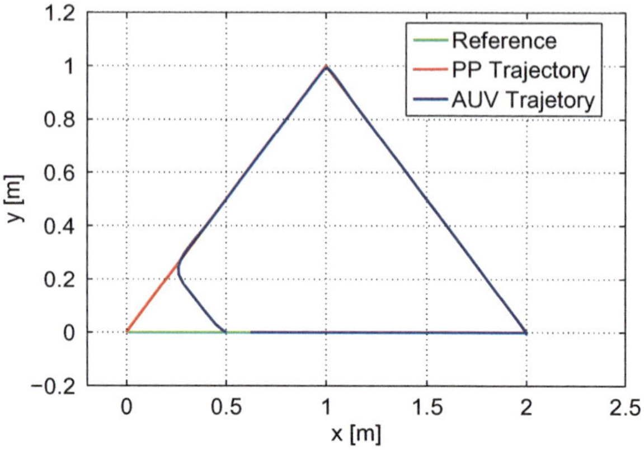

Fig. 5.10 PF results with nondifferentiable path

Notice that in Sect. 5.2.1, the desired path is assumed to be smooth. This is to facilitate the unified expression of parametrization of the ZPE manifold in (5.4)–(5.10). For a nondifferentiable desired path such as the following example:

$$p(s) = \begin{cases} p_x(s) = s, \quad p_y(s) = s & s \in [0, 1), \\ p_x(s) = s, \quad p_y(s) = 2 - s & s \in [1, 2), \\ p_x(s) = 4 - s, \quad p_y(s) = 0 & s \in [2, 4), \end{cases}$$

we need to specify the $[p_\psi(s), p_u(s), p_v(s), p_r(s)]$ at the nondifferentiable points. In the example, we adopt the following choice:

$$p(s) = \begin{cases} p_\psi(s) = \pi/4, \quad p_u(s) = \sqrt{2}\dot{s}, \quad p_v(s) = 0, \quad p_r(s) = 0, & s = 0, \\ p_\psi(s) = -\pi/4, \quad p_u(s) = \sqrt{2}\dot{s}, \quad p_v(s) = 0, \quad p_r(s) = 0, & s = 1, \\ p_\psi(s) = -\pi/2, \quad p_u(s) = \dot{s}, \quad p_v(s) = 0, \quad p_r(s) = 0, & s = 2. \end{cases}$$

In the simulation, the AUV, subject to 20% model error, is expected to follow the above nondifferentiable path; the speed assignment to be $u_r = 0.2$ (m/s). As the simulation results (Figs. 5.10, 5.11, and 5.12) indicate, the proposed MOMPC method can well handle the nondifferentiable path and the parametric uncertainty. Furthermore, we can observe obvious decrease in surge velocity at these nondifferentiable points. The decrease in surge velocity conforms with the objective prioritization we set for the path-following control and also largely increases the control stability.

5.6 Notes and Summary

In this chapter, the path-following control problem was studied for the autonomous underwater vehicles. To handle the prioritization between PF requirements, we have proposed a novel MOMPC method supported by two implementation algorithms:

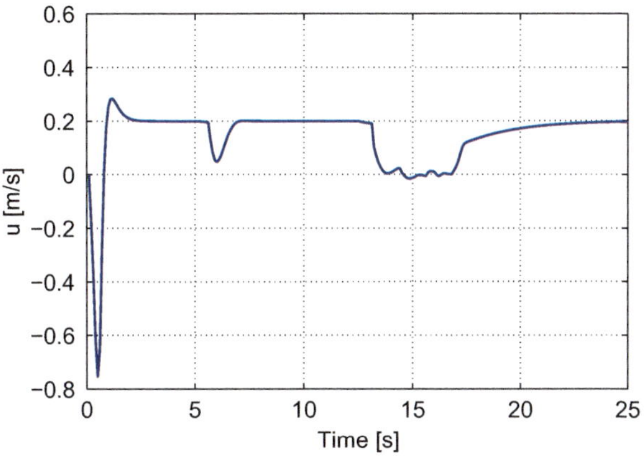

Fig. 5.11 Surge velocity of the AUV (nondifferentiable)

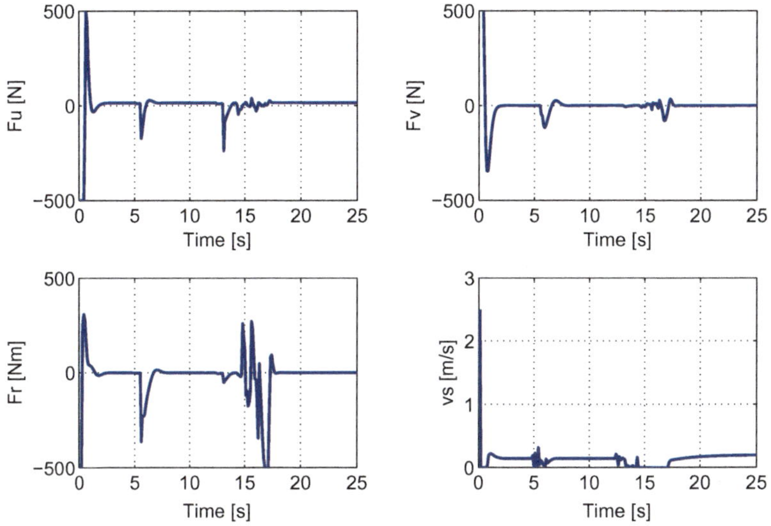

Fig. 5.12 Control inputs of the augmented system (nondifferentiable)

the WS-based and the LO-based algorithms. An explicit parametrization of the zero-path-error manifold was constructed facilitating the MOMPC formulation for the AUV PF control problem. For the WS-MOMPC scheme, a logistic function was proposed to automate the weight selection for the objectives. The PMP was applied with several well-defined barrier functions. The sufficient conditions that guarantee the convergence of the closed-loop trajectory were derived. The internal relationship between the WS-MOMPC and the LO-MOMPC was explored. Simulation results on the Falcon dynamic model demonstrated the effectiveness of the proposed method: The objective prioritization could be well incorporated into the PF control. Furthermore, the robustness test results indicated that the MOMPC method had good inherent robustness, which presented a promising, useful, and powerful framework for the AUVs to solve a broad spectrum of motion control problems.

References

Alessandretti A, Aguiar AP, Jones CN (2013) Trajectory-tracking and path-following controllers for constrained underactuated vehicles using model predictive control. In: Proceedings of the 2013 European control conference (ECC). IEEE, pp 1371–1376

Bemporad A, de la Peña DM (2009) Multiobjective model predictive control. Automatica 45(12):2823–2830

Böck M, Kugi A (2013) Real-time nonlinear model predictive path-following control of a laboratory tower crane. IEEE Trans Control Syst Technol 22(4):1461–1473

Boyd S, Boyd SP, Vandenberghe L (2004) Convex optimization. Cambridge University Press, Cambridge

Conn AR, Gould NI, Toint PL (2000) Trust region methods. SIAM

Deb K (2011) Multi-objective optimisation using evolutionary algorithms: an introduction. In: Multi-objective evolutionary optimisation for product design and manufacturing. Springer, pp 3–34

Diehl M, Ferreau HJ, Haverbeke N (2009) Efficient numerical methods for nonlinear MPC and moving horizon estimation. In: Nonlinear model predictive control. Springer, pp 391–417

Encarnação P, Pascoal A, Arcak M (2000) Path following for marine vehicles in the presence of unknown currents. IFAC Proc 33(27):507–512

Encarnacao P, Pascoal A (2000) 3D path following for autonomous underwater vehicle. In: Proceedings of the 39th IEEE conference on decision and control (CDC), vol. 3. IEEE, pp 2977–2982

Faulwasser T (2013) Optimization-based solutions to constrained trajectory-tracking and path-following problems. Ph.D. thesis, Otto-von-Guericke-Universitat, Magdeburg, Germany, 2012

Faulwasser T, Findeisen R (2010) Constrained output path-following for nonlinear systems using predictive control. IFAC Proc 43(14):753–758

Faulwasser T, Kern B, Findeisen R (2009) Model predictive path-following for constrained nonlinear systems. In: Proceedings of the 48h IEEE conference on decision and control (CDC) held jointly with 2009 28th Chinese control conference. IEEE, pp 8642–8647

Faulwasser T, Weber T, Zometa P, Findeisen R (2016) Implementation of nonlinear model predictive path-following control for an industrial robot. IEEE Trans Control Syst Technol 25(4):1505–1511

Graichen K (2012) A fixed-point iteration scheme for real-time model predictive control. Automatica 48(7):1300–1305

He D, Wang L, Sun J (2015) On stability of multiobjective NMPC with objective prioritization. Automatica 57:189–198

Kerrigan EC, Maciejowski JM (2002) Designing model predictive controllers with prioritised constraints and objectives. In: Proceedings of 2002 IEEE international symposium on computer aided control system design. IEEE, pp 33–38

Lapierre L, Jouvencel B (2008) Robust nonlinear path-following control of an AUV. IEEE J Oceanic Eng 33(2):89–102

Lapierre L, Soetanto D (2007) Nonlinear path-following control of an AUV. Ocean Eng 34(11–12):1734–1744

Mayne DQ (2014) Model predictive control: recent developments and future promise. Automatica 50(12):2967–2986

Micaelli A, Samson C (1993) Trajectory tracking for unicycle-type and two-steering-wheels mobile robots. Research Report RR-2097, INRIA

Miettinen K (2012) Nonlinear multiobjective optimization, vol. 12. Springer

Müller MA, Allgöwer F (2012) Improving performance in model predictive control: switching cost functionals under average dwell-time. Automatica 48(2):402–409

Ocampo-Martinez C, Ingimundarson A, Puig V, Quevedo J (2007) Objective prioritization using lexicographic minimizers for MPC of sewer networks. IEEE Trans Control Syst Technol 16(1):113–121

Oh SR, Sun J (2010) Path following of underactuated marine surface vessels using line-of-sight based model predictive control. Ocean Eng 37(2–3):289–295

Ohtsuka T (2004) A continuation/GMRES method for fast computation of nonlinear receding horizon control. Automatica 40(4):563–574

Pavlov A, Nordahl H, Breivik M (2009) MPC-based optimal path following for underactuated vessels. IFAC Proc 42(18):340–345

Qin SJ, Badgwell TA (2000) An overview of nonlinear model predictive control applications. In: Nonlinear model predictive control, pp 369–392

Rawlings JB, Mayne DQ, Diehl M (2017) Model predictive control: theory, computation, and design. Nob Hill Publishing

Shen C, Buckham B, Shi Y (2017) Modified C/GMRES algorithm for fast nonlinear model predictive tracking control of AUVs. IEEE Trans Control Syst Technol 25(5):1896–1904

Shen C, Shi Y, Buckham B (2017) Integrated path planning and tracking control of an AUV: a unified receding horizon optimization approach. IEEE/ASME Trans Mechatron 22(3):1163–1173

Yu S, Li X, Chen H, Allgöwer F (2015) Nonlinear model predictive control for path following problems. Int J Robust Nonlinear Control 25(8):1168–1182

Zambrano D, Camacho EF (2002) Application of MPC with multiple objective for a solar refrigeration plant. In: Proceedings of the international conference on control applications, vol 2. IEEE, pp 1230–1235

Zavala VM, Flores-Tlacuahuac A (2012) Stability of multiobjective predictive control: a utopia-tracking approach. Automatica 48(10):2627–2632

Chapter 6
Distributed Lyapunov-Based Model Predictive Formation-Tracking Control for AUVs Subject to Disturbances

6.1 Introduction

6.1.1 Research Background and Contributions

In Chaps. 3–5, we have discussed MPC for the trajectory tracking, dynamic positioning, and path-following problems for a single AUV. In this chapter, we will see how the MPC can be applied to solve cooperative control problems among multiple AUVs. The design procedure is exemplified by the distributed predictive formation-tracking control of a multi-AUV system which requires the AUVs to keep a prescribed formation while tracking a given reference trajectory simultaneously.

Most of existing control methods for the multi-AUV system cannot achieve the optimal control performance with various constraints such as actuator saturation or safety constraint. In contrast, MPC is known as one of the few control methods that can explicitly handle system constraints and optimize the control performance (Mayne et al. 2000; Chen and Allgöwer 1998; Liu et al. 2018). Problems concerning trajectory tracking or formation tracking with MPC strategies have been extensively studied, e.g., Gu and Hu (2006); Shen et al. (2018); Dunbar and Murray (2006); Wang and Ding (2014). The authors propose an MPC approach for the regulation and tracking problem of the nonholonomic mobile robot in Gu and Hu (2006). The stability of the closed-loop system is guaranteed by a local feedback controller and terminal ingredients. As reported in Wang and Ding (2014), the formation tracking together with the collision avoidance of a group of nonholonomic mobile robots is realized by an DMPC strategy. In Li et al. (2017), a DMPC-based formation-tracking method is designed for the multi-AUV system without disturbances, where only the kinematic properties of AUVs are considered. The main challenge of applying MPC lies in the guarantee of the closed-loop stability. Conventional MPC usually designs the terminal cost function, the terminal set and terminal control law to ensure the closed-loop stability (Chen and Allgöwer 1998). However, these stability analysis techniques are practically inefficient for AUVs due to the complex nonlinear system dynamics. One promising solution is to circumvent the local linearization by employing the LMPC

Y. Shi et al., *Advanced Model Predictive Control for Autonomous Marine Vehicles*, Advances in Industrial Control, https://doi.org/10.1007/978-3-031-19354-5_6

(de la Peña and Christofides 2008; Shen et al. 2018; Liu et al. 2017). LMPC combines the merits of both the Lyapunov-based control and the MPC, making the MPC design constructive and intuitive. More importantly, the LMPC method inherits the stability and robustness of the Lyapunov-based controller, which is attractive for the formation-tracking controller design. Therefore, we propose a DLMPC approach for the formation-tracking problem of multi-AUV systems.

Furthermore, the collision avoidance is considered for multi-AUV systems to ensure the safe operation. Several effective collision avoidance methods have been developed [see, e.g., Panagou et al. (2016), Wang and Ding (2014), Wen et al. (2017)]. In Panagou et al. (2016), a novel class of Lyapunov barrier functions is introduced to achieve the collision avoidance objective. As stated in Wang and Ding (2014), MPC is applied to solve this problem by imposing the collision avoidance constraint into the optimization problem. In this chapter, an additionally potential field-based cost term, which accounts for the collision avoidance, is designed and incorporated into the DLMPC optimization problem.

In practice, AUVs are always exposed to unknown external environmental disturbances induced by the ocean current, which poses a significant challenge to the formation control task (Chen et al. 2013). For this reason, the formation control that ensures the robustness against the disturbances becomes necessary. Robust MPC methods have been applied to systems with uncertainties in different forms such as tube-based MPC and min-max MPC (Mayne et al. 2005; Limon et al. 2006). It is worth noting that stochastic disturbances can also be tackled within the paradigm of MPC (Chatterjee and Lygeros 2015). Compared to robust MPC, the ESO-based method is more constructive and computationally tractable regarding the complex dynamics. As such, we further incorporate the ESO design into the DLMPC scheme, where the ESO can estimate the unknown ocean current disturbances.

The main contributions of this chapter are summarized as follows:

- A distributed model predictive formation-tracking control scheme is proposed for multi-AUV systems, and the ESO is designed to deal with the unknown ocean current disturbances. To achieve the inter-AUV collision avoidance, a potential field-based collision cost term is designed. The control input for each AUV is determined by solving the local DMPC optimization problem considering its own and the neighbors' state information. The overall control performance and robustness of the formation tracking are improved.
- The recursive feasibility of the DLMPC algorithm and the closed-loop stability of multi-AUV systems are ensured by the stability constraint that is constructed based on the ESO-based auxiliary controller and its associated Lyapunov function. The theoretical results provide a firm ground for the formation-tracking controller design for practical multi-AUV systems.

6.1.2 Chapter Organization

The remainder part of this chapter is organized as follows: The problem formulation and preliminaries are given in Sect. 6.2. In Sect. 6.3, the main results are introduced including the collision avoidance cost design, the ESO-based auxiliary design, the DLMPC algorithm and the theoretical analysis of the feasibility and the stability. Section 6.4 provides the simulation results before the conclusion is provided in Sect. 6.5.

The infinity norm and the 2 norm are denoted by $\|\cdot\|_\infty$ and $\|\cdot\|$, respectively. $\|x\|_P^2$ denotes the weighted Euclidean norm $x^{\mathrm{T}}Px$, where P is a positive definite matrix, the column operation $[x_1^{\mathrm{T}}, x_2^{\mathrm{T}}, \ldots, x_n^{\mathrm{T}}]^{\mathrm{T}}$ is written as $\mathrm{col}(x_1, x_2, \ldots, x_n)$. The superscript 'T' represents the transposition operation; $\mathrm{diag}(\cdot)$ denotes the diagonal operation. $\lambda_{\min}(P)$ and $\lambda_{\max}(P)$ denote the minimum and maximum eigenvalues of P, respectively.

6.2 Problem Statement and Preliminaries

6.2.1 AUV Modeling

In this chapter, the 3-DOF horizontal motion of the AUV is considered under a reasonable assumption that the roll angle ϕ and the pitch angle θ are small. Neglecting the elements corresponding to heave, roll, and pitch, a group of AUVs with homogeneous dynamics is considered. The kinematics of AUV i is described by

$$\dot{\eta}_i = R(\psi_i)\nu_i, \tag{6.1}$$

where $\eta_i = [x_i, y_i, \psi_i]^{\mathrm{T}}$ includes the position vector $p_i = [x_i, y_i]^{\mathrm{T}}$ and the orientation ψ_i, and $\nu_i = [u_i, v_i, r_i]^{\mathrm{T}}$ denotes the velocity vector. $R(\psi_i)$ is the rotation matrix in yaw

$$R(\psi_i) = \begin{bmatrix} \cos\psi_i & -\sin\psi_i & 0 \\ \sin\psi_i & \cos\psi_i & 0 \\ 0 & 0 & 1 \end{bmatrix}.$$

The 3-DOF nonlinear dynamic motion equation of AUV i is expressed as

$$M_i\dot{\nu}_i + C_i(\nu_i)\nu_i + D_i(\nu_i)\nu_i + g_i(\eta_i) = \tau_i + \tau_{iw}, \tag{6.2}$$

where M_i is the system inertia matrix including the added mass of AUV i, $D_i(\nu_i)$ denotes the damping matrix, the gravitational and buoyancy forces are assumed to outbalance each other such that the restoring force $g_i(\eta_i) = 0$, τ_i denotes the control input, $\tau_{iw} = M_i R^{\mathrm{T}}(\psi_i)w_i$, w_i denotes the ocean current disturbances, and $C_i(\nu_i)$ denotes the Coriolis-centripetal matrix including added mass.

Combining (6.1) and (6.2), we establish the system model for the tracking problem of the AUV i,

$$\dot{x}_i = \begin{bmatrix} R(\psi_i)\nu_i \\ M_i^{-1}(\tau_i + \tau_{iw} - C_i(\nu_i)\nu_i - D_i(\nu_i)\nu_i) \end{bmatrix} = f(x_i, \tau_i, w_i), \quad (6.3)$$

where $x_i = \text{col}(\eta_i, \nu_i) \in \mathbb{R}^6$ and $\tau_i \in \mathbb{R}^3$ are the state and the control input of the AUV i, respectively. The following important properties of the AUV model are easily explored and will be exploited in the controller design:

- P-1: The inertia matrix is symmetric, positive definite, and upper bounded, i.e., $M_i = M_i^{\text{T}} \succ 0$, $\|M_i\|_\infty = \bar{m}_i$, where $\bar{m}_i$ is a known constant.
- P-2: The Coriolis matrix is skew-symmetric, i.e., $C_i(\nu_i) = -C_i^{\text{T}}(\nu_i)$.
- P-3: The rotation matrix satisfies $R^{-1}(\psi_i) = R^{\text{T}}(\psi_i)$ and $\|R^{\text{T}}(\psi_i)\eta_i\| = \|\eta_i\|$.
- P-4: The damping matrix is positive definite, i.e., $D_i(\nu_i) \succ 0$.

6.2.2 *Problem Formulation*

A directed graph $\mathcal{G} = (\mathcal{V}, \mathcal{E})$ is introduced to describe the interaction among AUVs, with the set of nodes $\mathcal{V} = \{1, \ldots, N\}$ and the set of all edges $\mathcal{E} = \{(i, j) \subset \mathcal{V} \times \mathcal{V}\}$. The edge (i, j) represents the communication link from the node i to the node j. The adjacency matrix $\mathcal{A} = [a_{ij}] \in \mathbb{R}^{N \times N}$ is associated with the graph $\mathcal{G}$. $a_{ij} = 1$ if $(i, j) \in \mathcal{E}$, i.e., the information can be exchanged between AUV i and j, and $a_{ij} = 0$ otherwise. The node i of the graph represents the associating AUV i.

Assumption 6.1 For multi-AUV systems, AUV $i, i \in \mathcal{V}$ could receive the information from the virtual leader and its neighbors. The AUVs are synchronized in time clock, and the system states are available at each sampling time instant t_k.

We now formulate the formation-tracking problem for AUVs. In order to maintain a prescribed formation shape and track a time-varying reference trajectory of the virtual leader $\eta_r(t) = \text{col}(p_r(t), \psi_r(t))$, AUV i is steered to satisfy:

(1) *Tracking:* $\lim_{t\to\infty}\{p_i(t) - p_r(t)\} = d_{ir}$, $\lim_{t\to\infty}\{\psi_i(t) - \psi_r(t)\} = 0$, where d_{ir} is the configuration formation vector, $j \in \mathcal{V}, i \neq j$ and $p_r = [x_r, y_r]^{\text{T}}$ is the reference position of the virtual leader;
(2) *Formation:* $\lim_{t\to\infty}\{p_i(t) - p_j(t)\} = d_{ij}$, where d_{ir} is the configuration formation vector, $j \in \mathcal{V}, i \neq j$;
(3) *Collision avoidance:* $\|p_i(t) - p_j(t)\| \geq 2R$, where R is the safety radius for each AUV i, $\lambda_i = d_{ir}$ denotes relative deviation.

To avoid singularities in the reference trajectory, we make the following assumption:

Assumption 6.2 The reference trajectory and its first-, second- and third-order derivatives are smooth and bounded, i.e., $0 \leq \underline{p} \leq \|p_r\|_\infty \leq \overline{p} < \infty$, $0 \leq \underline{p_1} \leq \|\dot{p}_r\|_\infty \leq \overline{p}_1 < \infty$, $0 \leq \underline{p_2} \leq \|\ddot{p}_r\|_\infty \leq \overline{p}_2 < \infty$, and $0 \leq \underline{p_3} \leq \|\dddot{p}_r\|_\infty \leq \overline{p}_3 < \infty$.

From the reference trajectory η_r, we make the following reference augmentation:

$$\begin{aligned}
\psi_r &= \text{atan2}(\dot{y}_r(t), \dot{x}_r(t)), \\
u_r &= \sqrt{\dot{x}_r^2(t) + \dot{y}_r^2(t)}, \\
v_r &= 0, \\
r_r &= (\dot{x}_r(t)\ddot{y}_r(t) - \dot{y}_r(t)\ddot{x}_r(t))/(\dot{x}_r^2(t) + \dot{y}_r^2(t)),
\end{aligned}$$

where atan2 is the four-quadrant inverse tangent operator.

6.3 Distributed Lyapunov-Based Model Predictive Formation-Tracking Control

In this section, an ESO-based auxiliary controller is firstly designed using the backstepping technique. Next, we propose distributed model predictive formation-tracking controller for the multi-AUV. Finally, the recursive feasibility and the stability of the multi-AUV system are analyzed.

6.3.1 *Design of the Auxiliary Control Law*

Here we construct the auxiliary control law using the backstepping technique. Let $\eta_i = [x_i, y_i, \psi_i]^{\mathrm{T}}$ denote the trajectory of AUV i, $i \in \mathcal{V}$ and $\eta_{ir} = \eta_r - \lambda_i = [x_{ir}, y_{ir}, \psi_{ir}]^{\mathrm{T}}$ be the desired path. Define the virtual reference trajectory

$$\begin{aligned}
\dot{\eta}_{id} &= \dot{\eta}_{ir} - \Lambda_i \tilde{\eta}_i, \\
\nu_{id} &= R^{-1}(\psi_i)\dot{\eta}_{id},
\end{aligned} \tag{6.4}$$

where $\tilde{\eta}_i = \eta_i - \eta_{ir}$ represents the tracking error of AUV i, Λ_i is a diagonal design matrix. Let

$$s_i = \dot{\eta}_i - \dot{\eta}_{id} = \dot{\tilde{\eta}}_i + \Lambda_i \tilde{\eta}_i, \tag{6.5}$$

where s_i is a new state variable.

The position error dynamics becomes

$$\begin{aligned}
\dot{\tilde{\eta}}_i &= \dot{\eta}_i - \dot{\eta}_{ir} \\
&= R(\psi_i)(\nu_i - \nu_{ir}) \\
&= R(\psi_i)(R^{-1}(\psi_i)s_i + \alpha_i - \nu_{ir}) \\
&= -\Lambda_i \tilde{\eta}_i + s_i,
\end{aligned} \tag{6.6}$$

in which $\alpha_i = \nu_{id} = R^{-1}(\psi_i)(\dot{\eta}_{ir} - \Lambda_i \tilde{\eta}_i)$ is a stabilizing function. Furthermore, we define

$$V_{i,1} = \frac{1}{2}\tilde{\eta}_i^{\mathrm{T}} K_{pi} \tilde{\eta}_i, \tag{6.7}$$

where $K_{pi} = \mathrm{diag}(k_{pi1}, k_{pi2}, k_{pi3})$ with $k_{pi1}, k_{pi2}, k_{pi3}$ being positive. Then, the derivative of $V_{i,1}$ becomes

$$\dot{V}_{i,1} = \tilde{\eta}_i^{\mathrm{T}} K_{pi} \dot{\tilde{\eta}}_i = (-\tilde{\eta}_i^{\mathrm{T}} K_{pi} \Lambda_i \tilde{\eta}_i + s_i^{\mathrm{T}} K_{pi} \tilde{\eta}_i). \tag{6.8}$$

The dynamics of AUV i can be expressed as

$$M_i^* \ddot{\eta}_i + C_i^* \dot{\eta}_i + D_i^* \dot{\eta}_i = R(\psi_i)(\tau_i + \tau_{iw}), \tag{6.9}$$

where $M_i^* = R(\psi_i) M_i R^{\mathrm{T}}(\psi_i)$, $C_i^* = R(\psi_i)[C_i(\nu_i) - M_i S(\dot{\psi}_i)]R^{\mathrm{T}}(\psi_i)$, $D_i^* = R(\psi_i)$ $D_i(\nu_i) R^{\mathrm{T}}(\psi_i)$ and

$$S(\dot{\psi}_i) = \begin{bmatrix} 0 & -\dot{\psi}_i & 0 \\ \dot{\psi}_i & 0 & 0 \\ 0 & 0 & 0 \end{bmatrix}.$$

In the next step, the Lyapunov function is defined as

$$V_{i,2} = V_{i,1} + \frac{1}{2} s_i^{\mathrm{T}} M_i^* s_i. \tag{6.10}$$

Taking the time derivative of $V_{i,2}$ results in

$$\begin{aligned}
\dot{V}_{i,2} =& \dot{V}_{i,1} + (s_i^{\mathrm{T}} M_i^* \dot{s}_i + \frac{1}{2} s_i^{\mathrm{T}} \dot{M}_i^* s_i) \\
=& s_i^{\mathrm{T}} (\frac{1}{2} \dot{M}_i^* - C_i^* - D_i^*) s_i + s_i^{\mathrm{T}} R(\psi_i)[\tau_i + \tau_{iw} - M_i \dot{\nu}_{id} - C_i(\nu_i)\nu_{id} - D_i(\nu_i)\nu_{id}] \\
& - \tilde{\eta}_i^{\mathrm{T}} K_{pi} \Lambda_i \tilde{\eta}_i + s_i^{\mathrm{T}} K_{pi} \tilde{\eta}_i \\
=& s_i^{\mathrm{T}} R(\psi_i)[\tau_i + \tau_{iw} - M_i \dot{\nu}_{id} - C_i(\nu_i)\nu_{id} - D_i(\nu_i)\nu_{id} + R(\psi_i)^{\mathrm{T}} K_{pi} \tilde{\eta}_i] - \tilde{\eta}_i^{\mathrm{T}} K_{pi} \Lambda_i \tilde{\eta}_i - s_i^{\mathrm{T}} D_i^* s_i,
\end{aligned} \tag{6.11}$$

where $s_i^{\mathrm{T}}(\dot{M}_i^* - 2C_i^*)s_i = 0$ since the property P-2 holds.

The ideal auxiliary control law is then designed as

$$\tau_i^a = M_i \dot{\nu}_{id} + C_i(\nu_i)\nu_{id} + D_i(\nu_i)\nu_{id} - R^{\mathrm{T}}(\psi_i) K_{pi} \tilde{\eta}_i - R^{\mathrm{T}}(\psi_i) K_{di} s_i - \tau_{iw}, \tag{6.12}$$

where $K_{di} = \mathrm{diag}(k_{di1}, k_{di2}, k_{di3})$ with $k_{di1}, k_{di2}, k_{di3}$ being positive. In practice, the accurate ocean current disturbances are not available. Since the ocean current has finite energy, we make the following assumption:

Assumption 6.3 The ocean current disturbances w_i and its rate $\dot{w}_i$ are unknown but bounded by $\|w_i\| \le \bar{w}_i$ and $\|\dot{w}_i\| \le \bar{\dot{w}}_i$, respectively.

In what follows, an ESO is designed to estimate the unknown disturbances (Peng and Wang 2018; Fu and Yu 2018). To facilitate the ESO design, define $\mu_i = R(\psi_i)\nu_i$, then the system in (6.3) is transformed to

$$\begin{aligned}\dot{\eta}_i &= \mu_i,\\ \dot{\mu}_i &= R(\psi_i)M_i^{-1}\tau_i + \phi_i(\eta_i, \mu_i) + w_i,\end{aligned} \tag{6.13}$$

where $\phi_i(\eta_i, \mu_i) = s(\dot{\psi}_i)\mu_i - R(\psi_i)M_i^{-1}[C(\nu_i) + D(\nu_i)]R^{\mathrm{T}}(\psi_i)\mu_i$.

Let $\hat{\eta}_i$, $\hat{\mu}_i$, and $\hat{w}_i$ be the estimation of η_i, μ_i and w_i, respectively. Define the estimation errors $e_{1i} = \eta_i - \hat{\eta}_i$, $e_{2i} = \mu_i - \hat{\mu}_i$, $e_{3i} = w_i - \hat{w}_i$. Then, the ESO for the AUV $i, i \in \mathcal{V}$ is designed as

$$\begin{cases}\dot{\hat{\eta}}_i = \hat{\mu}_i + K_1 e_{1i},\\ \dot{\hat{\mu}}_i = R(\psi_i)M_i^{-1}\tau_i + \phi_i(\eta_i, \mu_i) + K_2 e_{1i} + \hat{w}_i,\\ \dot{\hat{w}}_i = K_3 e_{1i},\end{cases} \tag{6.14}$$

where the matrices $K_1 = \mathrm{diag}(k_1, k_1, k_1)$, $K_2 = \mathrm{diag}(k_2, k_2, k_2)$, and $K_3 = \mathrm{diag}(k_3, k_3, k_3)$ are positive.

The estimation errors then become

$$\begin{cases}\dot{e}_{1i} = e_{2i} - K_1 e_{1i},\\ \dot{e}_{2i} = e_{3i} - K_2 e_{1i},\\ \dot{e}_{3i} = \dot{w}_i - K_3 e_{1i}.\end{cases} \tag{6.15}$$

Let $e_i = \mathrm{col}(e_{1i}, e_{2i}, e_{3i}) \in \mathbb{R}^9$, the estimation error dynamics (6.15) is expressed as

$$\dot{e}_i = Ae_i + B\dot{w}_i, \tag{6.16}$$

where

$$A = \begin{bmatrix}-K_1 & I_3 & 0_3\\ -K_2 & 0_3 & I_3\\ -K_3 & 0_3 & 0_3\end{bmatrix}, \quad B = \begin{bmatrix}0_3\\ 0_3\\ I_3\end{bmatrix}.$$

The stability of error dynamic system in (6.16) is established if the following inequality:

$$A^{\mathrm{T}}P + PA \leq -\varrho I_9, \tag{6.17}$$

where $P \succ 0$, $\varrho \geq 1$.

Based on the ESO (6.14), the auxiliary control law becomes

$$\hat{\tau}_i^a = M_i\dot{\nu}_{id} + C_i(\nu_i)\nu_{id} + D_i(\nu_i)\nu_{id} - R^{\mathrm{T}}(\psi_i)K_{pi}\tilde{\eta}_i - R^{\mathrm{T}}(\psi_i)K_{di}s_i - \hat{\tau}_{iw}. \tag{6.18}$$

The following lemma presents the stability of the multi-AUV system based on the auxiliary control law:

Lemma 6.1 *Consider a team of AUVs described by* (6.3), *the auxiliary control law* (6.18), *together with the ESO* (6.14). *Given a positive constant* ϱ, *if there exists a matrix* P *satisfying* (6.17) *and the following two conditions: (1)* $\varrho - \theta_1 - \theta_2 > 0$ *and (2)* $\lambda_{\min}(K_{di}^*) - \frac{\bar{m}_i^2}{2\theta_1} > 0$ *hold, with* $0 < \theta_1 < 1, 0 < \theta_2 < 1$. *Then, the multi-AUV systems achieve the formation-tracking task.*

Proof The Lyapunov candidate function of the ESO can be defined as

$$V_{i,3} = \frac{1}{2} e_i^{\mathrm{T}} P e_i, \tag{6.19}$$

which is bounded by

$$\frac{1}{2}\lambda_{\min}(P)\|e_i\|^2 \leq V_{i,3} \leq \frac{1}{2}\lambda_{\max}(P)\|e_i\|^2.$$

The derivative of $V_{i,3}$ with respect to (6.16) satisfies

$$\dot{V}_{i,3} = \frac{1}{2} e_i^{\mathrm{T}} (A^{\mathrm{T}} P + PA) e_i + e_i^{\mathrm{T}} P B \dot{w}_i. \tag{6.20}$$

Consider the following Lyapunov function:

$$V_i = V_{i,2} + V_{i,3}, \; V = \sum_{i=1}^{N} V_i. \tag{6.21}$$

Note that the Lyapunov function (6.21) is continuously differentiable and radically unbounded. By the Lyapunov theorem (Khalil 1996), there exist functions $\beta_i(\cdot), i = 1, 2$ that belong to class $\mathcal{K}_\infty$ such that $\beta_1(\|\mathbf{x}_i\|) \leq V_i \leq \beta_2(\|\mathbf{x}_i\|)$, with $\mathbf{x}_i = \mathrm{col}(\tilde{\eta}_i, s_i, e_i)$.

Under the ESO-based auxiliary control law (6.18), the derivative of the overall Lyapunov function becomes

$$\begin{aligned}
\dot{V}|_{\hat{\tau}_i^a(t)} &= \sum_{i=1}^{N} \Big\{ - \tilde{\eta}_i^{\mathrm{T}} K_{pi} \Lambda_i \tilde{\eta}_i - s_i^{\mathrm{T}} K_{di}^* s_i + s_i^{\mathrm{T}} M_i^* e_{3i} + \frac{1}{2} e_i^{\mathrm{T}} (A^{\mathrm{T}} P + PA) e_i + e_i^{\mathrm{T}} P B \dot{w}_i \Big\} \\
&\leq \sum_{i=1}^{N} \Big\{ - \tilde{\eta}_i^{\mathrm{T}} K_{pi} \Lambda_i \tilde{\eta}_i - (\lambda_{\min}(K_{di}^*) - \frac{\bar{m}_i^2}{2\theta_1}) \|s_i\|^2 - \frac{1}{2}(\varrho - \theta_1)\|e_i\|^2 + \epsilon_1 \Big\} \\
&= \sum_{i=1}^{N} \Big\{ - \tilde{\eta}_i^{\mathrm{T}} K_{pi} \Lambda_i \tilde{\eta}_i - (\lambda_{\min}(K_{di}^*) - \frac{\bar{m}_i^2}{2\theta_1}) \|s_i\|^2 - \frac{1}{2}(\varrho - \theta_1 - \theta_2)\|e_i\|^2 - \frac{1}{2}\theta_2 \|e_i\|^2 + \epsilon_1 \Big\},
\end{aligned} \tag{6.22}$$

where $\dot{V}|_{\hat{\tau}_i^a(t)}$ denotes the derivative of Lyapunov function under the control law $\hat{\tau}_i^a(t)$, $K_{di}^* = D_i^* + K_{di}$, $\epsilon_1 = \|e_i\| \|PB\| \bar{\dot{w}}_i$. If $\|e_i\| \geq 2\|PB\|\bar{\dot{w}}_i/\theta_2$ and the condition 1) and 2) hold, then

$$\dot{V}_{i,3} \leq -\frac{1}{2}(\varrho - \theta_1 - \theta_2)\|e_i\|^2 < 0. \tag{6.23}$$

Hence, the estimation errors converge to the compact set $\Omega_e = \{e_i \in \mathbb{R}^3 | \|e_i\| \leq \rho_{e_i}, \rho_{e_i} > 2\|PB\|\bar{w}_i/\theta_2\}$. In addition, $\dot{V}_i|_{\hat{\tau}_i^a(t)} < 0$, i.e., $\dot{V} < 0$, if the following inequality holds: $\|e_i\| > 2\|PB\|\bar{w}_i/\theta_2$. Then, the multi-AUV systems fulfill the formation-tracking task. This completes the proof. □

6.3.2 Design of the Optimization Problem

6.3.2.1 Collision Avoidance Design

The potential function is widely used in robotics for the collision avoidance (Shim et al. 2003; Mastellone et al. 2008). The potential function is designed such that the closer distance between AUVs has the higher potential value, otherwise the AUV has the lower potential value. The collision avoidance cost term is designed by using the potential field function that penalizes collisions between different AUVs. The collision cost $J_{ij}^{ca}(t)$ can be expressed as

$$J_{ij}^{ca}(t) = \sum_{j=1}^{N} a_{ij} \frac{\Gamma}{1 + \exp(d_{ij}^{ca})} = \sum_{j=1}^{N} a_{ij} f_{pi}(t), \tag{6.24}$$

where $d_{ij}^{ca} = k_i(\|\tilde{D}_{ij}\| - 2R)$, $\|\tilde{D}_{ij}\|$ is the distance between AUV i and j, $f_{pi}(t)$ is the collision avoidance function, $\Gamma > 0$ is a tuning parameter, $k_i > 0$ is a parameter that defines the smoothness of the collision avoidance function. By setting appropriate parameters of the collision avoidance function, AUVs can avoid collisions under the optimal solution.

6.3.2.2 Cost Function Design

For each AUV i, $i \in \mathcal{V}$, the cost function at time t_k is defined as

$$\begin{aligned} & J_i(x_i(t_k), \tilde{x}_j^a(s; t_k), \tau_i(s; t_k)) \\ &= \int_{t_k}^{t_k+T} \left\{ \sum_{j=1}^{N} \left[a_{ij} \|x_{ij}(s; t_k)\|_{Q_i}^2 + \|x_{ir}(s; t_k)\|_{Q_i'}^2 \right] + J_{ij}^{ca}(\tilde{P}_i(s), \tilde{P}_j^a(s)) + \|\tau_i(s; t_k)\|_{R_i}^2 \right\} \mathrm{d}s, \end{aligned} \tag{6.25}$$

where $s \in [t_k, t_k + T]$, T denotes the prediction horizon, δ denotes the sampling period, i.e., $t_{k+1} = t_k + \delta$, $Q_i \succeq 0$, $Q_i' \succeq 0$, $R_i \succ 0$ are the weighting matrices, $x_{ir} = \tilde{x}_i + \Lambda_i - x_r$, $x_r = \mathrm{col}(\eta_r, \nu_r)$, $x_{ij} = \tilde{x}_i + \Lambda_i - \tilde{x}_j^a - \Lambda_j$. $\Lambda_i = \mathrm{col}(\lambda_i, \dot{\lambda}_i)$.

For simplicity, the collision avoidance cost $J_{ij}^{ca}(\tilde{P}_i(s), \tilde{P}_j^a(s))$ is abbreviated as $J_{ij}^{ca}(s)$, which is introduced in Sect. 6.3.2.1. $\tilde{P}_i$ is the nominal position of AUV i and $\tilde{P}_j^a$ is the assumed position of AUV j. $\tilde{x}_i$ is the nominal state of AUV i, which is evaluated by $\dot{\tilde{x}}_i(s; t_k) = f(\tilde{x}_i(s; t_k), \tau_i(s; t_k), 0)$. The assumed state $\tilde{x}_j^a(s; t_k)$ of AUV j at t_k is generated by $\dot{\tilde{x}}_j^a(s; t_k) = f(\tilde{x}_j^a(s; t_k), \tilde{\tau}_j^a(s; t_k), 0)$, and the assumed control input is $\tilde{\tau}_j^a(s; t_k)$, i.e.,

$$\tilde{\tau}_j^a(s; t_k) = \begin{cases} \tau_j^*(s; t_{k-1}), s \in [t_k, t_{k-1} + T), \\ \tau_j^*(t_{k-1} + T; t_{k-1}), s \in [t_{k-1} + T, t_k + T]. \end{cases}$$

6.3.2.3 The Optimization Problem

For each AUV i, $i \in \mathcal{V}$, using the objective function (6.25), the optimization problem $\mathcal{P}_i$ is designed as,

$$\min_{\tau_i \in S(\delta)} J_i(x_i(t_k), \tilde{x}_j^a(s; t_k), \tau_i(s; t_k)) \tag{6.26a}$$

$$\text{s.t. } \dot{\tilde{x}}_i(s; t_k) = f(\tilde{x}_i(s; t_k), \tau_i(s; t_k), 0), \tag{6.26b}$$

$$\tilde{x}_i(t_k; t_k) = x_i(t_k), \tag{6.26c}$$

$$\|\tau_i(s; t_k)\|_\infty \leq \tau^{\max}, \tag{6.26d}$$

$$\dot{V}_i|_{\tau_i(t_k)} \leq \dot{V}_i|_{\hat{\tau}_i^a(t_k)}, \tag{6.26e}$$

where $S(\delta)$ is a family of piecewise constant function with the sampling period δ, $s \in [t_k, t_k + T]$, $\tau_i^*(t_k)$ are the control input and optimal control input for the AUV i at t_k, respectively. $\hat{\tau}_{iw}(t_k)$ is the estimated disturbance, $\hat{\tau}_i^a(t_k)$ is the ESO-based auxiliary control law. In addition, $\tau_i^*(s; t_k)$ denotes the optimal solution of the optimization problem $\mathcal{P}_i$. $x_i(t_k)$ and $x_i^*(s; t_k)$ denote the actual state and the optimal state of the AUV i, respectively. Furthermore, based on (6.11), (6.18) and the derivative of the associated Lyapunov function, we can acquire the detailed expression of the stability constraint (6.26e) at time instant t_k as follows:

$$\begin{aligned} & s_i^{\mathrm{T}} R(\psi_i)[\tau_i(t_k) - M_i\dot{v}_{id} - C_i(v_i)v_{id} - D_i(v_i)v_{id} + R(\psi_i)^{\mathrm{T}} K_{pi}\tilde{\eta}_i] - \tilde{\eta}_i^{\mathrm{T}} K_{pi}\Lambda_i\tilde{\eta}_i - s_i^{\mathrm{T}} D_i^* s_i \\ & \leq - s_i^{\mathrm{T}} R(\psi_i)\hat{\tau}_{iw} - \tilde{\eta}_i^{\mathrm{T}} K_{pi}\Lambda_i\tilde{\eta}_i - s_i^{\mathrm{T}} K_{di}^* s_i, \end{aligned} \tag{6.27}$$

where $\tau_i(t_k)$ represents the optimization variable.

Remark 6.1 It is worth noting that (6.26b) is the nominal model of AUV i, $i \in \mathcal{V}$, which can be used to predict the state trajectory. (6.26c) is the initial state condition at t_k. (6.26d) specifies the control input constraint. (6.26e) is the stability constraint that is constructed by the ESO-based auxiliary controller $\hat{\tau}_i^a(t_k)$ as well as the associated Lyapunov function V_i. The DLMPC controller inherits the stability and robustness of the ESO-based auxiliary controller (Liu et al. 2010).

Remark 6.2 Commonly used robust MPC techniques, including tube and min-max MPC have a high requirement of the computation resource. In the DLMPC scheme, where the ocean current disturbances are estimated by the ESO, is computationally efficient and will not introduce additional computational burden.

6.3.3 Feasibility and Stability Analysis

To make the DLMPC strategy applicable, it is necessary to ensure that the formation-tracking optimization problem $\mathcal{P}_i$ admits a solution at each time instant. To guarantee the feasibility of the multi-AUV formation-tracking optimization problem, the following lemmas are introduced:

Lemma 6.2 *Suppose that each AUV i is controlled by the auxiliary control input* (6.18)*, then the Coriolis matrix $C_i(v_i)$ and the damping matrix $D_i(v_i)$ are bounded, i.e., $\|C_i(v_i)\|_\infty \leq \bar{c}_i$, $\|D_i(v_i)\|_\infty \leq \bar{d}_i$.*

Proof The proof can be obtained by following the similar line of Shen et al. (2018) [Lemma 1]. □

Lemma 6.3 *Let $\bar{K}_p$ and $\bar{K}_d$ denote the largest entity in control gains $K_{pi}\Lambda_i$ and K_{di}, respectively. If the following condition can be ensured, i.e.:*

$$\bar{m}_i\bar{\dot{v}}_{id} + (\bar{c}_i + \bar{d}_i)\bar{v}_{id} + (\bar{K}_p + \bar{K}_d)\|\Gamma_i(t_0)\| + \bar{m}_i(\bar{\hat{w}}_i) \leq \tau^{\max},$$

then the auxiliary control input $\|\hat{\tau}_i^a(t)\|_\infty \leq \tau^{\max}$.

Proof According to the definition, $|\psi_r| \leq \pi$. As $\dot{\psi}_r = |r_r| \leq \bar{p}_1\bar{p}_2/(\underline{p}_1)^2 = \bar{\dot{\psi}}_r$. Here $\bar{\dot{\psi}}_r$ represents the maximum value of $\dot{\psi}_r$, and other variables are defined in the similar way in the sequel. Then, $\ddot{\psi}_r$ can be expressed as

$$\begin{aligned}|\ddot{\psi}_r| =& \left|\frac{\dot{x}_r\dddot{y}_r - \dot{y}_r\dddot{x}_r}{\dot{x}_r^2 + \dot{y}_r^2} - \frac{2(\dot{x}_r\ddot{y}_r - \dot{y}_r\ddot{x}_r)(\dot{x}_r\ddot{x}_r + \dot{y}_r\ddot{y}_r)}{(\dot{x}_r^2 + \dot{y}_r^2)^2}\right| \\ \leq& \bar{p}_1\bar{p}_3/\underline{p}_1^2 + 2\bar{p}_1^2\bar{p}_2^2/\underline{p}_1^2 = \bar{\ddot{\psi}}_r,\end{aligned}$$

where $0 < \bar{\ddot{\psi}}_r < \infty$. So we can acquire

$$\begin{aligned}\|\eta_r\|_\infty &= \bar{\eta}_r = \max\{\overline{p}, \pi\}, \\ \|\dot{\eta}_r\|_\infty &= \bar{\dot{\eta}}_r = \max\{\overline{p}_1, \bar{\dot{\psi}}_r\}, \\ \|\ddot{\eta}_r\|_\infty &= \bar{\ddot{\eta}}_r = \max\{\overline{p}_2, \bar{\ddot{\psi}}_r\}.\end{aligned}$$

Let $\Gamma_i = \text{col}(\tilde{\eta}_i, R^{\mathrm{T}}(\psi_i)s_i, e_i)$, the Lyapunov function (6.21) can be rewritten as $V_i = \frac{1}{2}\Gamma_i^{\mathrm{T}}s_i\Gamma_i$, where $s_i = \text{diag}(K_p, M_i, P)$. As Lemma 6.1 holds, $\dot{V}_i < 0$,

$\|\Gamma_i\| \leq \|\Gamma_i(t_0)\|$. Furthermore, $\|\tilde{\eta}_i\|_\infty \leq \|\tilde{\eta}_i\| \leq \|\Gamma_i\|$, $\|s_i\|_\infty \leq \|s_i\| \leq \|\Gamma_i\|$, and $\|e_i\|_\infty \leq \|e_i\| \leq \|\Gamma_i\|$, then $\|\tilde{\eta}_i\|_\infty \leq \|\Gamma_i(t_0)\|$, $\|s_i\|_\infty \leq \|\Gamma_i(t_0)\|$, and $\|e_i\|_\infty \leq \|\Gamma_i(t_0)\|$. Hence,

$$\begin{aligned}\|\dot{\tilde{\eta}}_i\|_\infty &= \|s_i - \tilde{\eta}_i\|_\infty \\ &\leq \|s_i\|_\infty + \|\tilde{\eta}_i\|_\infty \\ &\leq 2\|\Gamma_i(t_0)\|.\end{aligned} \tag{6.28}$$

Note that

$$\begin{aligned}\dot{\eta}_{id} &= \dot{\eta}_{ir} - \Lambda_i\tilde{\eta}_i, \\ \nu_{id} &= R^{\mathrm{T}}(\psi_i)\dot{\eta}_{id}.\end{aligned} \tag{6.29}$$

Then, we have

$$\dot{\nu}_{id} = R^{\mathrm{T}}(\psi_i)\ddot{\eta}_{id} + R(\psi_i)s(\dot{\psi}_i)\dot{\eta}_{id}. \tag{6.30}$$

As $\|s_i\|_\infty$ and $\|\tilde{\eta}_i\|_\infty$ are bounded, (6.5) and (6.28) imply $\dot{\eta}_{id}$, $\ddot{\eta}_{id}$, and $\dot{\tilde{\eta}}_i$ are bounded, i.e., $\|\dot{\eta}_{id}\|_\infty \leq \bar{\dot{\eta}}_{id}$ and $\|\ddot{\eta}_{id}\|_\infty \leq \bar{\ddot{\eta}}_{id}$. Consequently, the upper bounds of the $\|\nu_{id}\|_\infty$ and $\|\dot{\nu}_{id}\|_\infty$ can be acquired

$$\|\nu_{id}\|_\infty \leq \bar{\nu}_{id}, \|\dot{\nu}_{id}\|_\infty \leq \bar{\dot{\nu}}_{id}. \tag{6.31}$$

Hence,

$$\begin{aligned}&\|\hat{\tau}_i^a(t)\|_\infty \\ \leq&\|M_i\dot{\nu}_{id}\|_\infty + \|(C_i(\nu_i) + D_i(\nu_i))\nu_{id}\|_\infty + \|R^{\mathrm{T}}(\psi_i)(K_{pi}\Lambda_i\tilde{\eta}_i + K_{di}s_i)\|_\infty + \|\hat{\tau}_{iw}\|_\infty \\ \leq&\bar{m}_i\|\dot{\nu}_{id}\|_\infty + (\bar{c}_i + \bar{d}_i)\|\nu_{id}\|_\infty + \bar{K}_p\|\tilde{\eta}_i\|_\infty + \bar{K}_d\|s_i\|_\infty + \|\hat{\tau}_{iw}\|_\infty \\ \leq&\bar{m}_i\bar{\dot{\nu}}_{id} + (\bar{c}_i + \bar{d}_i)\bar{\nu}_{id} + \bar{K}_p\|\Gamma_i(t_0)\| + \bar{K}_d\|\Gamma_i(t_0)\| + \bar{m}_i\bar{\hat{w}}_i \\ \leq&\tau^{\max},\end{aligned}$$

where $\bar{\hat{w}}_i = \bar{w}_i + \|\Gamma_i(t_0)\|$, which concludes the proof. □

In what follows, the recursive feasibility and stability of the multi-AUV system under the DLMPC algorithm are presented in Theorem 6.1.

Theorem 6.1 *For the multi-AUV system described by* (6.3) *with unknown ocean current disturbances, each AUV i solves the formation-tracking optimization problem $\mathcal{P}_i$ at each sampling time instant t_k. If there exists a feasible solution at the initial instant t_0, then the proposed method is recursively feasible, i.e., the optimization problem $\mathcal{P}_i$ admits a solution for all $t_k \geq t_0$. The formation tracking of the multi-AUV system can be achieved under the control input generated by the proposed method.*

Proof We first establish the recursive feasibility of the proposed method before the the stability proof. We prove the feasibility using mathematical induction. In order to guarantee the recursive feasibility of the optimization problem $\mathcal{P}_i$ of each AUV i, $i \in \mathcal{V}$, we need to find an input sequence that satisfies the constraint (6.26d) and (6.26e). Note that the auxiliary control input $\hat{\tau}_i^a(t_{k+1})$ is generated by the auxiliary controller for the multi-AUV system at the time instant t_{k+1}. Therefore, the constraint (6.26e) is satisfied if we choose the control input as the auxiliary control law, i.e., $\tau_i(t_{k+1}; t_{k+1}) = \hat{\tau}_i^a(t_{k+1})$. If the optimization problem $\mathcal{P}_i$ is feasible at t_k, and the optimal solution is $\tau_i^*(s; t_k)$, $s \in [t_k, t_k + T]$. At time instant t_{k+1}, the feasible control input can be chosen as

$$\tau_i(s; t_{k+1}) = \begin{cases} \hat{\tau}_i^a(t_{k+1}), s \in [t_{k+1}, t_{k+1} + \delta), \\ \tau_i^*(s; t_k), s \in [t_{k+1} + \delta, t_k + T), \\ \tau_i^*(t_k + T; t_k), s \in [t_k + T, t_{k+1} + T). \end{cases}$$

Also, due to the results in Lemma 6.3, thus, the control input constraint (6.26d) is satisfied. Therefore, the recursive feasibility can be guaranteed.

Since Lemma 6.1 holds, we have

$$\dot{V}_i|_{\hat{\tau}_i^a(t_k)} < 0.$$

Considering (6.26e) and the optimal solution τ_i^*, to be implemented at each sampling instant, we have

$$\dot{V}_i|_{\tau_i^*(t_k)} \leq \dot{V}_i|_{\hat{\tau}_i^a(t_k)} < 0. \tag{6.32}$$

The formation tracking of the multi-AUV system under the control input generated by the proposed method can be achieved. The stability of the overall multi-AUV system can be ensured. This concludes the proof. □

6.4 Simulation Study

In this section, we set up two simulation scenarios to verify the validity of the DLMPC algorithm for the multi-AUV system. Inter-AUV collision avoidance is tested in the first scenario with and without the external disturbances. In the second scenario, we test the disturbance rejection by introducing considerable disturbances. The simulation results illustrate the excellent formation-tracking performance, collision avoidance function, and the robustness of the DLMPC method.

6.4.1 Simulation Setup

The reference trajectory of the virtual leader is a sinusoidal shape trajectory and defined as follows:

$$x_r = 0.5t,$$
$$y_r = \sin(0.5t + \frac{\pi}{2}). \tag{6.33}$$

In the simulation, the time-varying disturbances induced by ocean current are simulated by

$$w_i = \begin{bmatrix} 0.5v_i^3 + 0.3\sin(0.7t) \\ 0.6u_i r_i + 0.3u_i + 0.2\sin(0.6t) \\ -0.5v_i^2 - u_i r_i - 0.2\sin(0.9t) \end{bmatrix}. \tag{6.34}$$

The dynamic model parameters of Falcon are as follows (Shen et al. 2017): $m = 116(\text{kg})$ and the moment of inertia with respect to the z-axis $I_z = 13.1(\text{kg m}^2)$. The hydrodynamic coefficients $X_{\dot{u}} = -167.6$, $Y_{\dot{v}} = -477.2$, $N_{\dot{r}} = -15.9$, $X_u = 26.9$, $Y_v = 35.8$, $N_r = 3.5$, $D_u = 241.3$, $D_v = 503.8$, and $D_r = 76.9$. $M_x = m - X_{\dot{u}}$, $M_y = m - Y_{\dot{v}}$, and $M_\psi = I_z - N_{\dot{r}}$, $D_i(v_i)$=diag(X_u, Y_v, N_r)+diag$(D_u|u_i|, D_v|v_i|, D_r|r_i|)$. $M_i = \text{diag}(M_x, M_y, M_\psi)$. $C_i(v_i) = [0, 0, -M_y v_i; 0, 0, M_x u_i; M_y v_i, -M_x u_i, 0]$.

The parameters of DLMPC are designed as follows: The sampling period $\delta = 0.1$s, the prediction horizon $T = 6\delta$. The safety radius $R = 0.25$m. The tuning parameter $\Gamma = 10^4$ and the smoothness parameter $k_i = 8$. The weighting matrices are chosen as $Q_i = \text{diag}(10^4, 10^4, 10^3, 10, 10, 10)$, $Q_i' = \text{diag}(10^4, 10^4, 10^3, 10^2, 10^2, 10^2)$, and $R_i = \text{diag}(10^{-3}, 10^{-3}, 10^{-3})$. The parameters of the ESO are selected as $K_1 = \text{diag}(5, 5, 5)$, $K_2 = \text{diag}(6, 6, 6)$, and $K_3 = \text{diag}(8, 8, 8)$. The limit on control input is 1000N (Nm). The desired formation and tracking vectors are $D_{1r} = [0, 1]^{\text{T}}$, $D_{2r} = [0, -1]^{\text{T}}$, $D_{3r} = [1, 0]^{\text{T}}$, $D_{12} = -D_{21} = [0, 2]^{\text{T}}$, $D_{13} = -D_{31} = [1, 1]^{\text{T}}$, $D_{23} = -D_{32} = [1, -1]^{\text{T}}$. $\mathcal{N}_1 = \{0, 3\}$, $\mathcal{N}_2 = \{0, 3\}$, $\mathcal{N}_3 = \{0, 1, 2\}$.

6.4.2 Formation Tracking with Collision Avoidance

In the first scenario, the initial conditions for the three AUVs are chosen as $[0, 2, \frac{\pi}{2}, 0.4, 0, 0]^{\text{T}}$, $[-0.5, -0.5, \frac{3\pi}{4}, 0.4, 0, 0]^{\text{T}}$, $[-0.1, 0.1, -\frac{\pi}{2}, 0.4, 0, 0]^{\text{T}}$, respectively. In the first two parts, collision avoidance is considered without the disturbances while in the third part AUVs are exposed to disturbances.

In the first part, formation-tracking task of AUVs is considered without the collision avoidance. The simulation results of DLMPC are illustrated in Figs. 6.1 and 6.2. The trajectories of AUVs in the X–Y plane are plotted in Fig. 6.1, and the dash circles represent the safety radius of AUVs. During the period of [0.6s, 0.7s], AUV2 and AUV3 collide. As seen in Fig. 6.1, three AUVs can track the reference trajectory with the predesigned shape, but cannot avoid the collision. The optimal control input is shown in Fig. 6.2. This simulation provides a benchmark for the collision avoidance test.

In the second part, the formation-tracking task is reconsidered with the collision avoidance and the results are illustrated in Figs. 6.3 and 6.4. The formation-tracking

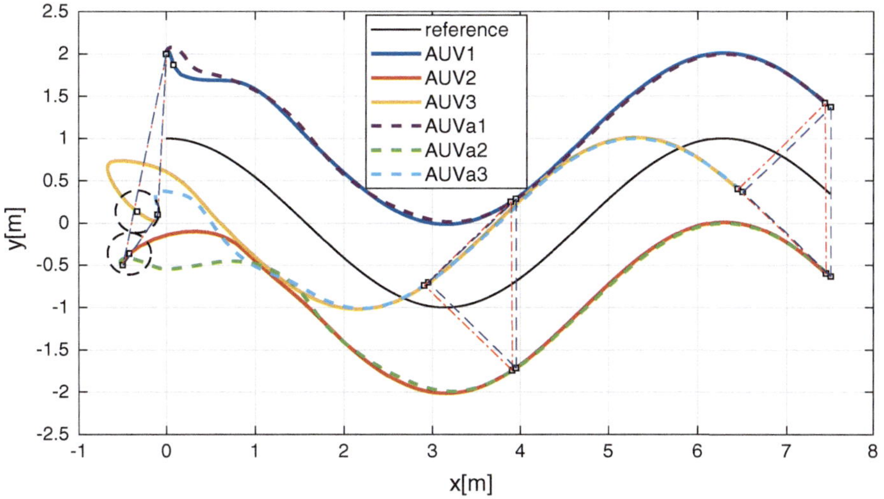

Fig. 6.1 Trajectories of three AUVs controlled by the DLMPC and the reference trajectory in the X–Y plane (without collision avoidance)

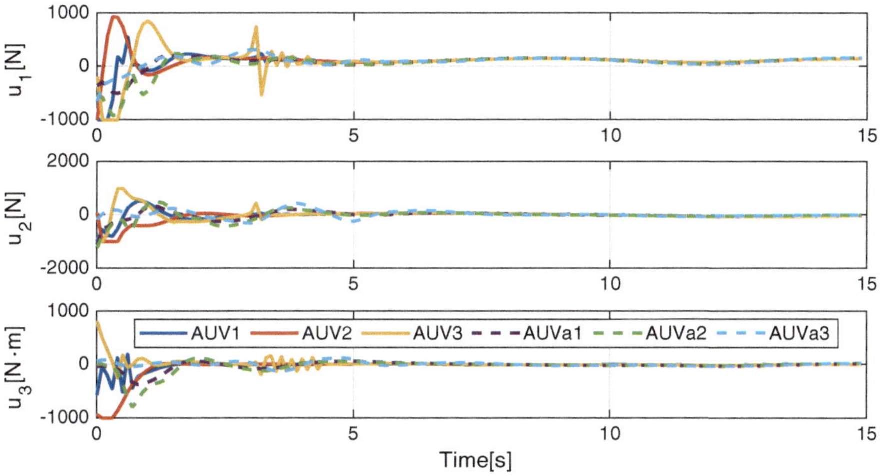

Fig. 6.2 Control inputs of three AUVs under the DLMPC

results under the auxiliary control law are also presented in Figs. 6.3 and 6.4. The trajectories of three AUVs are plotted in Fig. 6.3, in which the collision is avoided during the whole operational period. AUV1 denotes the first AUV controlled by DLMPC, and AUVa1 denotes the first AUV controlled by the auxiliary controller. It can be observed that three AUVs track the reference trajectory with the prescribed geometric shape. AUVs under DLMPC achieve faster and more accurate formation-tracking performance. Figure 6.4 depicts the optimal control input.

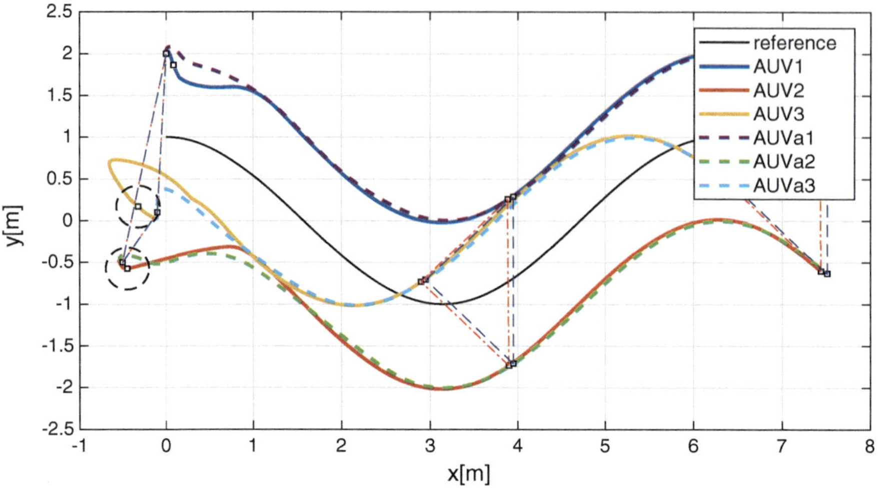

Fig. 6.3 Reference trajectory and the trajectories of AUVs with collision avoidance in the X–Y plane

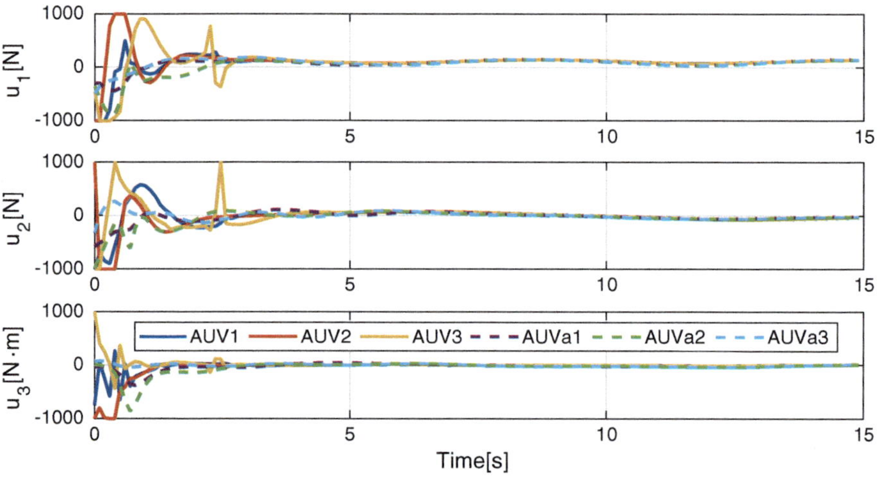

Fig. 6.4 Control input signals of three AUVs under the auxiliary control law and the DLMPC

6.4.3 *Multi-AUV Formation Tracking with Disturbances*

In the second scenario, AUVs are exposed to the unknown ocean current disturbances. The initial condition for the three AUVs are $[0, 2, \frac{\pi}{2}, 0.5, 0, 0]^{\mathrm{T}}$, $[-0.5, -0.5, \frac{3\pi}{4}, 0.5, 0, 0]^{\mathrm{T}}$, $[-1.5, 1.1, -\frac{\pi}{2}, 0.5, 0, 0]^{\mathrm{T}}$, respectively. The formation-tracking performance of the proposed algorithm and the auxiliary control law for the multi-AUV are illus-

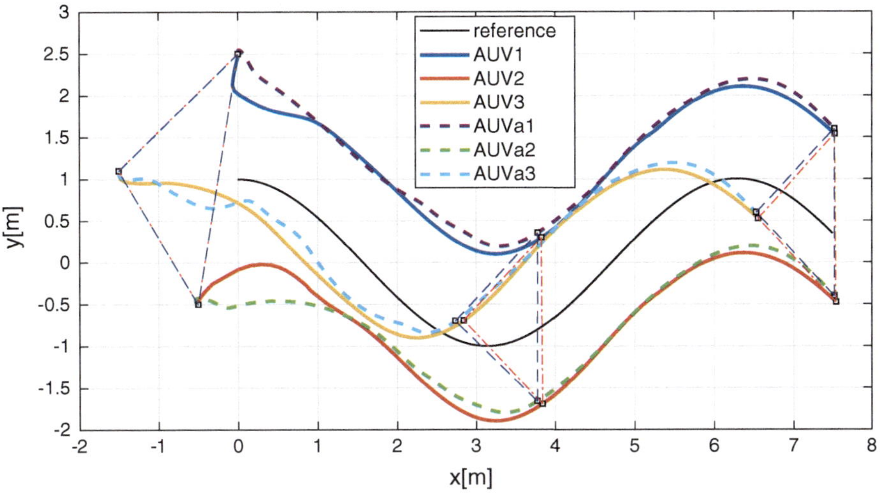

Fig. 6.5 Trajectories of three AUVs and reference trajectory in X–Y plane with disturbances (with the ESO)

trated in Figs. 6.5, 6.6 and 6.7. The simulated disturbances follow (6.34) for each AUV i.

In the first part, the AUVs controlled by DLMPC and the auxiliary control law without the ESO cannot accomplish the formation-tracking task. In the second part, the result of AUVs controlled by the DLMPC and the ESO-based auxiliary control law are shown in Fig. 6.5. As shown in Fig. 6.5, AUVs under the DLMPC quickly converge to the reference trajectory with the predesigned formation. The control inputs obtained by DLMPC and the auxiliary controller are shown in Fig. 6.6. The trajectories of Lyapunov function with ESO-based auxiliary controller and DLMPC are illustrated in Fig. 6.7. It implies that (1) DLMPC converge faster than the auxiliary controller from the perspective of evolutions of Lyapunov function; (2) DLMPC improves the formation-tracking performance of the multi-AUV system.

Quantitatively, the mean square errors (MSEs) of the formation tracking using four different controllers are summarized in Table 6.1. The smaller MSEs value means better formation-tracking performance. As can be seen, the proposed DLMPC with ESO achieves the smallest MSE. The formation-tracking performance is improved using DLMPC with ESO compared with the performance under the auxiliary controllers without the ESO. If the AUVs are exposed to ocean current disturbances, the overall system stability cannot be ensured under the auxiliary controller (without ESO) and DLMPC (without ESO).

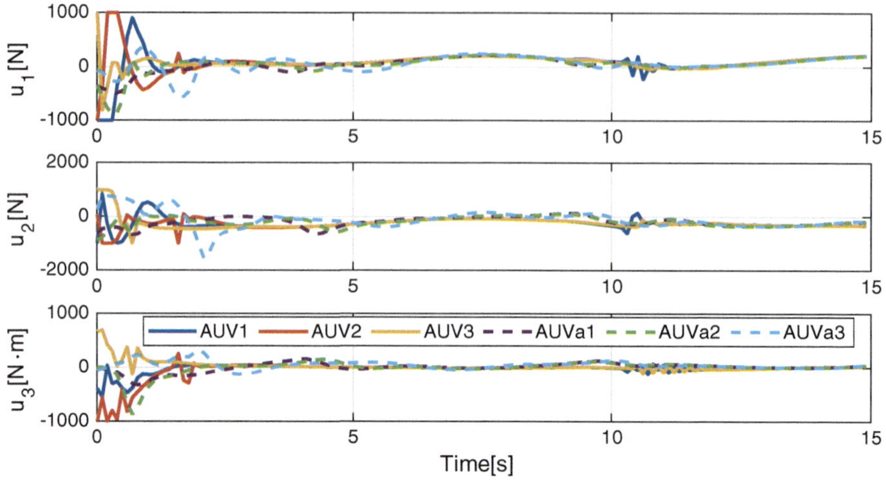

Fig. 6.6 Control inputs of AUVs with disturbances (with the ESO)

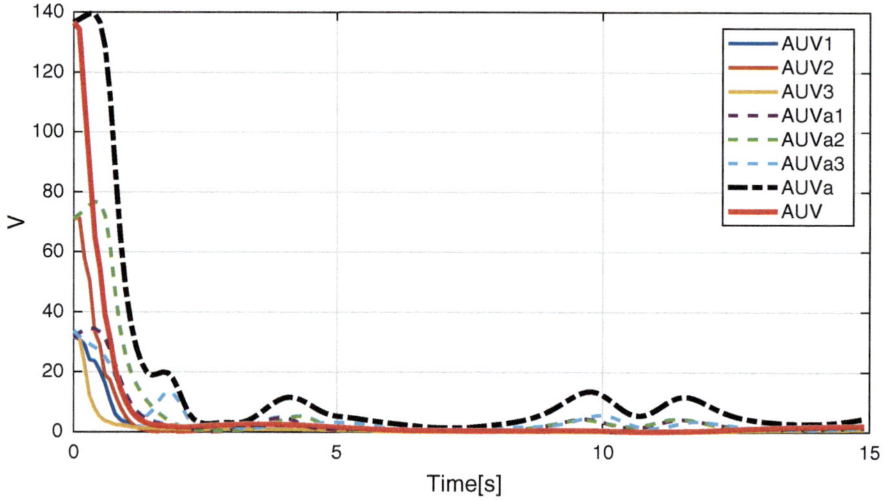

Fig. 6.7 Comparison of the Lyapunov function trajectories under ESO-based auxiliary control law and DLMPC

Table 6.1 Formation-tracking MSE of AUVs under DLMPC and auxiliary control law

Method	$x[\mathrm{m}^2]$	$y[\mathrm{m}^2]$	$\psi[\mathrm{rad}^2]$
Auxiliary controller (without ESO)	–	unstable	–
DLMPC (without ESO)	NaN	NaN	NaN
Auxiliary controller (with ESO)	0.1063	0.1235	1.2374
DLMPC (with ESO)	**0.0611**	**0.0551**	**0.3650**

6.5 Notes and Summary

In this chapter, we have proposed the DLMPC algorithm for the formation-tracking task of AUVs with unknown ocean current disturbances. The inter-AUV collision avoidance was achieved by designing artificial potential field-based cost term with well-tuned parameters, which was integrated with the formation-tracking cost function. By incorporating a stability constraint generated by the ESO-based auxiliary controller into the DMPC optimization problem, the control performance as well as the robustness was guaranteed. The recursive feasibility and closed-loop stability of the DLMPC control were rigorously proved. The simulation results verified the effectiveness of the proposed distributed Lyapunov-based model predictive formation-tracking control algorithm.

References

Chatterjee D, Lygeros J (2015) On stability and performance of stochastic predictive control techniques. IEEE Trans Autom Control 60(2):509–514

Chen H, Allgöwer F (1998) A quasi-infinite horizon nonlinear model predictive control scheme with guaranteed stability. Automatica 34(10):1205–1217

Chen M, Ge SS, How BEV, Choo YS (2013) Robust adaptive position mooring control for marine vessels. IEEE Trans Control Syst Technol 21(2):395–409

de la Peña DM, Christofides PD (2008) Lyapunov-based model predictive control of nonlinear systems subject to data losses. IEEE Trans Autom Control 53(9):2076–2089

Dunbar WB, Murray RM (2006) Distributed receding horizon control for multi-vehicle formation stabilization. Automatica 42(4):549–558

Fu M, Yu L (2018) Finite-time extended state observer-based distributed formation control for marine surface vehicles with input saturation and disturbances. Ocean Eng 159:219–227

Gu D, Hu H (2006) Receding horizon tracking control of wheeled mobile robots. IEEE Trans Control Syst Technol 14(4):743–749

Khalil HK (1996) Nonlinear systems, 2nd edn. Prentice Hall, New York

Li H, Xie P, Yan W (2017) Receding horizon formation tracking control of constrained underactuated autonomous underwater vehicles. IEEE Trans Ind Electron 64(6):5004–5013

Limón D, Alamo T, Salas F, Camacho EF (2006) Input to state stability of min-max MPC controllers for nonlinear systems with bounded uncertainties. Automatica 42(5):797–803

Liu C, Gao J, Li H, Xu D (2018) Aperiodic robust model predictive control for constrained continuous-time nonlinear systems: an event-triggered approach. IEEE Trans Cybern 48(5):1397–1405

Changxin Liu, Jian Gao, Demin Xu (2017) Lyapunov-based model predictive control for tracking of nonholonomic mobile robots under input constraints. Int J Control Autom Syst 15(5):2313–2319

Liu J, de la Peña DM, Christofides PD (2010) Distributed model predictive control of nonlinear systems subject to asynchronous and delayed measurements. Automatica 46(1):52–61

Mastellone S, Stipanović DM, Graunke CR, Intlekofer KA, Spong MW (2008) Formation control and collision avoidance for multi-agent non-holonomic systems: theory and experiments. Int J Robot Res 27(1):107–126

Mayne DQ, Rawlings JB, Rao CV, Scokaert POM (2000) Constrained model predictive control: Stability and optimality. Automatica 36(6):789–814

Mayne DQ, Seron MM, Raković SV (2005) Robust model predictive control of constrained linear systems with bounded disturbances. Automatica 41(2):219–224

Panagou D, Stipanović DM, Voulgaris PG (2016) Distributed coordination control for multi-robot networks using Lyapunov-like barrier functions. IEEE Trans Autom Control 61(3):617–632
Peng Z, Wang J (2018) Output-feedback path-following control of autonomous underwater vehicles based on an extended state observer and projection neural networks. IEEE Trans Syst Man Cybern Syst 48(4):535–544
Shen C, Shi Y, Buckham B (2017) Integrated path planning and tracking control of an AUV: a unified receding horizon optimization approach. IEEE/ASME Trans Mechatron 22(3):1163–1173
Shen C, Shi Y, Buckham B (2018) Trajectory tracking control of an autonomous underwater vehicle using Lyapunov-based model predictive control. IEEE Trans Ind Electron 65(7):5796–5805
Shim DH, Kim HJ, Sastry S (2003) Decentralized nonlinear model predictive control of multiple flying robots. In: Proceedings of 42nd IEEE international conference on decision and control (CDC), vol 4. IEEE, pp 3621–3626
Wang P, Ding B (2014) Distributed RHC for tracking and formation of nonholonomic multi-vehicle systems. IEEE Trans Autom Control 59(6):1439–1453
Wen G, Chen CLP, Liu Y-J (2017) Formation control with obstacle avoidance for a class of stochastic multiagent systems. IEEE Trans Ind Electron 65(7):5847–5855

Chapter 7
Robust Distributed Model Predictive Platooning Control for Heterogeneous Autonomous Surface Vehicles

7.1 Introduction

7.1.1 Research Background and Contributions

The platooning control of autonomous vehicles has attracted an increasing attention in the intelligent transportation area due to its potential to improve road capacity and efficiency (Guanetti et al. 2018; Feng et al. 2019). The platooning coordination of multi-vehicle, which requires vehicles to travel at the desired spacing and speed, further reduces the aerodynamic drag and improves the utilization of the road, especially for heavy-duty vehicles (Alam et al. 2015; Turri et al. 2016). In a similar vein, the platooning control for ASVs can also hugely improve waterborne transport efficiency and reduce labor costs (Viel et al. 2019). In practice, multi-vehicle systems are subject to various physical constraints, e.g., the control input and state constraints (Zheng et al. 2016), and external disturbances (Lin and Görges 2019). Therefore, the design of effective platooning control is one of the primary objectives for ASVs and autonomous vehicles study.

Recently, various notable control methods have been reported in the literature to tackle the autonomous vehicle platoon problem, such as optimal control (Jin and Orosz 2016), sliding mode control (Wu et al. 2020), delay-based spacing policy (Besselink and Johansson 2017). In comparison with the aforementioned control methods, DMPC gains attention from platooning control researchers since it is able to explicitly handle hard constraints and optimize the closed-loop performance. Also, DMPC is computationally efficient and applicable to dynamically decoupled vehicle platoon systems (Kianfar et al. 2015; Liu et al. 2018). In (Dunbar and Caveney 2012), the authors present a DMPC algorithm for the homogeneous nonlinear vehicle platoon and also derive a sufficient condition to ensure the string stability. In Li et al. (2016), each vehicle can communicate with its predecessor and follower. Based on the local and neighboring state information, the platooning control input for each agent is generated by solving a local DMPC optimization problem with a neighboring γ-gain stability constraint. The string stability constraints described above suppress the

Y. Shi et al., *Advanced Model Predictive Control for Autonomous Marine Vehicles*, Advances in Industrial Control, https://doi.org/10.1007/978-3-031-19354-5_7

spacing error at the price of a decreased feasibility. This may be conservative and result in unsatisfactory control performance. To achieve better control performance, the authors in Zheng et al. (2016) develop a DMPC algorithm without considering the string stability constraint for the heterogeneous autonomous vehicle platoon under four types of unidirectional communication topologies.

However, the methods in Dunbar and Caveney (2012), Li et al. (2016), Zheng et al. (2016) do not consider external disturbances, which may degrade the control performance, or destroy the stability of overall vehicle systems. Designing a robust controller that can deal with the disturbances is critically important to ensure the platooning task to be successful. In order to tackle this challenging issue, some robust DMPC methods have been developed. Tube-based MPC generally achieves a good tradeoff between the control performance and online computational requirement, and it has been insightfully studied in Mayne et al. (2005) for linear systems and in Mayne et al. (2011) for nonlinear systems. Further, tube-based decentralized MPC methods for general linear MASs (Trodden and Richards 2010) and general nonlinear MASs (Nikou and Dimarogonas 2019) have been developed. A recent paper (Feng et al. 2020) proposes a platooning control algorithm, which combines the tube-based DMPC with event-triggered feedforward control, for homogeneous linear autonomous vehicles. However, the overall system is assumed to be synchronized, which is hard to satisfy. In addition, Feng et al. (2020) does not consider the intervehicle safety, while it is necessary in safety-critical environment.

Inspired by the above discussions and existing problems, a nonlinear tube-based DMPC approach for the constrained heterogeneous ASV platoon is proposed. In the proposed algorithm, each ASV is assigned a local controller to calculate its local optimal control input, and then transmits its updated state trajectory to neighbors. The local system states and states of neighbors are required, when each ASV solves the local optimization problem. Since all ASVs optimize control inputs simultaneously, the updated optimal state trajectories cannot be received by its neighbors at the same time instant. Hence, the assumed state trajectory of each ASV, which is used as the optimal state trajectory, will be constructed and exchanged among ASVs. Similar to the methodology used to handle coupled constraints in Richards and How (2007), Trodden and Richards (2010), the neighbors' state trajectory is also estimated by each local controller. In this way, a coupled constraint is decoupled into a local constraint of the optimization problem. In particular, compared with Richards and How (2007); Trodden and Richards (2010), the transmission of assumed state trajectories among ASVs makes parallel execution of the proposed algorithm a reality.

The main contributions of this chapter are summarized as follows:

- A nonlinear robust DMPC approach is developed for the dynamically decoupled heterogeneous ASV platoon with the input constraint and bounded environmental disturbances. The novelties of the proposed distributed optimization problem lie in that: (a) Different from the method used to guarantee the intervehicle collision avoidance in Wei et al. (2021), a coupled minimum safe following distance constraint is designed and incorporated into the optimization problem to ensure the intervehicle safety of the ASV platoon; (b) a novel cost term is introduced with the purpose of suppressing the error between the predicted and assumed state of each ASV, which does not require the overall system to be synchronized.

- The auxiliary control law, being designed offline, reduces the deviation between the actual system state and the optimal nominal system state. The auxiliary control law design is more computationally efficient in comparison with Mayne et al. (2011), in which the auxiliary control law is determined by solving another MPC problem.
- Finally, the recursive feasibility of the robust distributed model predictive platooning control algorithm is analyzed, and the heterogeneous ASV platoon is proved to be input-to-state stable.

7.1.2 Chapter Organization

The remainder of this chapter is organized as follows: Sect. 7.2 formulates the platooning control problem. In Sect. 7.3, the auxiliary control law design and the robust DMPC algorithm are presented. The feasibility and stability are proved in Sect. 7.4. Section 7.5 gives the simulation results of the ASV platoon. The conclusion is given in Sect. 7.6.

In this chapter, the following notation is used: Let $\mathbb{R}$ and $\mathbb{N}$ denote the sets of real numbers and non-negative integers, respectively. For any vector $x \in \mathbb{R}^n$, $\|x\|$ is the Euclidean norm, $\|x\|_P = x^{\mathrm{T}}Px$ denotes the weighted Euclidean norm, where P is a positive definite matrix. P^{T} denotes the matrix transpose of P. $\lambda_{\min}(P)$ denotes the minimum eigenvalue of the matrix P. Given two sets S_1 and S_2, their Minkowski sum is defined as $S_1 \oplus S_2 := \{s_1 + s_2 | s_1 \in S_1, s_2 \in S_2\}$ and the Pontryagin difference is defined as $S_1 \ominus S_2 := \{s | s + s_2 \in S_1, s_2 \in S_2\}$. A fixed graph, denoted by $\mathcal{G} = (\mathcal{V}^0, \mathcal{E})$, is to describe the information exchange among ASVs, where $\mathcal{V}^0 = \{i \in \mathbb{N} | 0 \le i \le N\}$ is the set of the nodes representing a platoon of ASVs and $\mathcal{E} = \{(i, j) \subset \mathcal{V}^0 \times \mathcal{V}^0\}$ is the set of all directed edges denoting the communication from the ASV i to the ASV j. $\mathcal{N}_i := \{j | (i, j) \in \mathcal{E}\}$ denotes the set of ASV i's neighbors, i.e., ASV i receives the information from ASV j, $j \in \mathcal{N}_i$. $\mathcal{V} = \{i \in \mathbb{N} | 1 \le i \le N\}$ denotes all following ASVs.

7.2 Problem Setup

7.2.1 Autonomous Surface Vehicle Modeling

A group of $N + 1$ dynamically decoupled nonlinear heterogeneous ASVs are considered in this work. According to Fossen (2002), the model of ASV i is characterized by

$$\dot{\eta}_i = R(\psi_i)\nu_i, \tag{7.1a}$$
$$M_i\dot{\nu}_i + C_i(\nu_i)\nu_i + D_i(\nu_i)\nu_i = u_i + w_i, \tag{7.1b}$$

where $\eta_i = [x_i \ y_i \ \psi_i]^{\mathrm{T}}$ includes the surge displacement x_i, sway displacement y_i, and yaw angle ψ_i; the velocity $\nu_i = [\mu_i \ \upsilon_i \ r_i]^{\mathrm{T}}$ includes the surge velocity μ_i, sway velocity υ_i and yaw velocity r_i. $u_i = [u_{ix} \ u_{iy} \ u_{i\psi}]^{\mathrm{T}}$ is the control input and $w_i = [w_{ix} \ w_{iy} \ w_{i\psi}]^{\mathrm{T}}$ denotes the external disturbance. The rotation matrix in yaw $R(\psi_i)$ is

$$R(\psi_i) = \begin{bmatrix} \cos\psi_i & -\sin\psi_i & 0 \\ \sin\psi_i & \cos\psi_i & 0 \\ 0 & 0 & 1 \end{bmatrix}.$$

The inertia matrix including the added mass is defined as $M_i = \mathrm{diag}(M_{x_i}, M_{y_i}, M_{\psi_i})$ with $M_{x_i} = m_i - X_{\dot{\mu}_i}$, $M_{y_i} = m_i - Y_{\dot{\upsilon}_i}$, and $M_{\psi_i} = I_{z_i} - N_{\dot{r}_i}$. m_i and I_{z_i} are the ship mass and the moment of inertia about the z-axis of the fixed body frame, respectively. The hydrodynamic damping matrix including the drag force and vortex-induced force is

$$D_i(\nu_i) = \mathrm{diag}(-X_{\mu_i}, -Y_{\upsilon_i}, -N_{r_i}) + \mathrm{diag}(-X_{\mu_i|\mu_i|}|\mu_i|, -Y_{\upsilon_i|\upsilon_i|}|\upsilon_i|, -N_{r_i|r_i|}|r_i|),$$

and the Coriolis and centripetal matrix is

$$C_i(\nu_i) = \begin{bmatrix} 0 & 0 & -M_{y_i}\upsilon_i \\ 0 & 0 & M_{x_i}\mu_i \\ M_{y_i}\upsilon_i & -M_{x_i}\mu_i & 0 \end{bmatrix}.$$

7.2.2 ASV Platoon Modeling

In this chapter, the platooning control of heterogeneous ASVs is studied in the one-dimensional (x-direction) space. Due to the symmetric structure of ASVs, the resulting decoupled surge dynamics of ASV i is given by

$$\dot{x}_i = \mu_i, \tag{7.2a}$$
$$M_{x_i}\dot{\mu}_i - X_{\mu_i}\mu_i - X_{\mu_i|\mu_i|}|\mu_i|\mu_i = u_{ix} + w_{ix}. \tag{7.2b}$$

Define $q_i(t) = \left[x_i(t) \ \mu_i(t)\right]^{\mathrm{T}} \in \mathbb{R}^2$ as the actual system state. The predecessor–leader following communication topology proposed in Zheng et al. (2016) is adopted. Under this type of communication topology, ASV i can receive the information from its preceding ASV $i-1$ and the lead ASV 0. Hence, the neighboring set of ASV i is denoted by $\mathcal{N}_i = \{0, i-1\}$, $i \in \mathcal{V}$. The state of lead ASV 0 is denoted as $q_0(t)$, $t \geq 0$, and the following ASV i, $i \in \mathcal{V}$, will track the desired state,

$$q_i^{\text{des}}(t) = q_0(t) - \begin{bmatrix} id^x \\ 0 \end{bmatrix}, t \geq 0, \tag{7.3}$$

where d^x is a constant denoting the desired spacing between ASV i and its preceding ASV $i-1$. For ASV $i, i \in \mathcal{V}^0$, the error state $z_i(t) = \left[x_i^e(t)\ \mu_i^e(t)\right]^{\mathrm{T}}$ is given by

$$z_i(t) = q_i(t) - q_i^{\text{des}}(t), \tag{7.4}$$

where $x_i^e(t)$ and $\mu_i^e(t)$ denote the error position and error velocity, respectively. Taking the time derivative of (7.4), the error dynamics of ASV i can be described as

$$\begin{aligned} \dot{z}_i(t) =& f_i(z_i(t)) + B_i u_i(t) + B_i w_i(t) \\ =& \begin{bmatrix} \mu_i - \mu_0 \\ M_{x_i}^{-1}(X_{\mu_i}\mu_i + X_{\mu_i|\mu_i|}|\mu_i|\mu_i) - \dot{\mu}_0 \end{bmatrix} + B_i u_i(t) + B_i w_i(t), \end{aligned} \tag{7.5}$$

where $B_i = [0\ M_{x_i}^{-1}]^{\mathrm{T}}$. By a slight abuse of notation, $u_i(t)$ and $w_i(t)$ are used to denote $u_{ix}(t)$ and $w_{ix}(t)$ in the sequel. Here the control input $u_i(t)$ is subject to the control input constraint given by $u_i(t) \in \mathbb{U}_i \subseteq \mathbb{R}$. $w_i(t) \in \mathbb{W}_i := \{|w_i| \leq w_{i,\max}\} \subseteq \mathbb{R}$ represents the additive disturbance.

For the error system in (7.5), a standard assumption is given as in Dunbar and Caveney (2012).

Assumption 7.1 For ASV $i, i \in \mathcal{V}^0$, 1) The function $f_i(z_i) : \mathbb{R}^2 \to \mathbb{R}^2$ is continuously differentiable with $f_i(0) = 0$; 2) The system in (7.5) with any initial state $z_i(0)$ has a unique, absolutely continuous solution for the control input $u_i(t) \in \mathbb{U}_i$ and $w_i(t) \in \mathbb{W}_i, t \geq 0$.

Without considering the disturbances $w_i(t)$, the nominal error dynamics of the ASV i can be written as

$$\dot{\bar{z}}_i(t) = f_i(\bar{z}_i(t)) + B_i \bar{u}_i(t), \tag{7.6}$$

where $\bar{z}_i(t) \in \mathbb{R}^2$ and $\bar{u}_i(t) \in \mathbb{R}$ are the nominal error state and nominal control input, respectively. Next, the definition of input-to-state stable (ISS) is recalled (Jiang and Wang 2001).

Definition 7.1 Consider the system given by $\dot{x}(t) = f(x(t), w(t))$, where $x(t)$ is the system state, and $w(t)$ is the bounded input disturbance. Then, the system is ISS, if there exist a $\mathcal{KL}$ function $\alpha_1(\cdot,\cdot)$ and a $\mathcal{K}$ function $\gamma(\cdot)$, such that,

$$\|x(t)\| \leq \alpha_1(\|x(0)\|, t) + \gamma(\|\boldsymbol{w}\|_{[0,t]}),$$

where $\|\boldsymbol{w}\|_{[0,t]} := \sup\{\|w(s)\| | 0 \leq s \leq t\}$.

7.2.3 Control Objectives

The main control objective of this work is to realize platooning control of a group of heterogeneous ASVs subject to the input constraint and external disturbances. ASV $i, i \in \mathcal{V}$ maintains the desired spacing with its preceding ASV $i-1$ and tracks the speed of the lead ASV 0. Furthermore, in order to ensure the safety of the ASV platoon, the spacing between ASV i and ASV $i-1$ satisfies

$$x_i(t) - x_{i-1}(t) \leq -d_i^{\text{safe}}, \tag{7.7}$$

where $d_i^{\text{safe}} := (\mu_{i,\max} t_s + \mu_{i,\max}^2 / 2b_s)$ is the minimum safe following distance, t_s is the minimum reaction time, b_s is the maximum deceleration, and $\mu_{i,\max}$ is the maximum speed. From the definition of error state in (7.4), the inequality in (7.7) can be further written as

$$E(z_i(t) - z_{i-1}(t)) \leq -d_i^{\text{safe}} + d^x, \tag{7.8}$$

where $E = \begin{bmatrix} 1 & 0 \end{bmatrix}$, and $d_i^{\text{safe}} \leq d^x$. It is worth noting that since the predecessor–leader following communication topology is adopted in this chapter, ASV i only considers the minimum safe following distance d_i^{safe} between ASV i and its preceding ASV $i-1$.

In this work, a robust DMPC control method is proposed to achieve the control objective of the heterogeneous ASV platoon. The control input applied to ASV i consists of two components: (1) The nominal control action $\bar{u}_i^*(t)$, generated by solving the DMPC problem, aims to steer the nominal error state $\bar{z}_i(t)$ toward the origin, and (2) the auxiliary control law $\pi_i(z_i(t), \bar{z}_i(t))$, which is designed offline, targets at keeping the actual error state trajectory $z_i(t)$ in a specific set centered along the nominal error state trajectory $\bar{z}_i(t)$.

It is assumed that the DMPC problem for each ASV has the same prediction horizon T and sampling period $\delta \in (0, T]$. Let t_k be the sampling time instant, where $k \in \mathbb{N}$, and $t_{k+1} = t_k + \delta$. At t_k, define $\bar{z}_i(\cdot; t_k)$ as the nominal error state trajectory of ASV i, $\bar{z}_{i,i-1}(\cdot; t_k)$ as the nominal error state trajectory of ASV $i-1$ predicted by ASV i, $\bar{z}_i^*(\cdot; t_k)$ as the optimal nominal error state trajectory, $\bar{z}_{i,i-1}^*(\cdot; t_k)$ as the optimal nominal error state trajectory of ASV $i-1$ predicted by ASV i, $z_i^a(\cdot; t_k)$ as the assumed error state trajectory, and $\bar{z}_i^p(\cdot; t_k)$ as the predicted error state trajectory. The corresponding control inputs are denoted likewise.

ASV $i, i \in \mathcal{V}$ receives the assumed error state trajectory $z_{i-1}^a(s; t_k), s \in [t_k, t_k + T]$ from its preceding ASV $i-1$. The reason of implementing this information transmission strategy is that the optimal nominal error state trajectory $\bar{z}_{i-1}^*(s; t_k), s \in [t_k, t_k + T]$ of the preceding ASV $i-1$ is not available for its following ASV i at the same time instant t_k. Hence, the assumed nominal error state trajectory $z_{i-1}^a(s; t_k), s \in [t_k, t_k + T]$, based on the optimal nominal error states at t_{k-1}, is used to approximate the optimal nominal error state trajectory $\bar{z}_{i-1}^*(s; t_k), s \in [t_k, t_k + T]$. For ASV i, the assumed nominal error state trajectory $z_i^a(\cdot; t_k)$ at time t_k is

$$z_i^a(s; t_k) = \begin{cases} \bar{z}_i^*(s; t_{k-1}), & s \in [t_k, t_{k-1} + T], \\ 0, & s \in (t_{k-1} + T, t_k + T]. \end{cases} \tag{7.9}$$

Similarly, the assumed control $u_i^a(\cdot; t_k)$ is constructed as

$$u_i^a(s; t_k) = \begin{cases} \bar{u}_i^*(s; t_{k-1}), & s \in [t_k, t_{k-1} + T], \\ 0, & s \in (t_{k-1} + T, t_k + T]. \end{cases}$$

Also, ASV i will receive $z_{i-1}^a(\cdot; t_k)$ of ASV $i - 1$.

With this notation, the cost function of ASV i, $i \in \mathcal{V}^0$ at time t_k is defined as

$$\begin{aligned} & J_i\left(\bar{\mathbf{z}}_i(t_k), z_{i-1}^a(\cdot; t_k), \bar{\mathbf{u}}_i(\cdot; t_k)\right) \\ = & \int_{t_k}^{t_k+T} L_i\left(\bar{\mathbf{z}}_i(s; t_k), z_{i-1}^a(s; t_k), \bar{\mathbf{u}}_i(s; t_k)\right) \mathrm{d}s + F_i\left(\bar{\mathbf{z}}_i(t_k + T; t_k)\right), \end{aligned}$$

where $\bar{\mathbf{z}}_i(t_k) = [\bar{z}_i(t_k)\ \bar{z}_{i,i-1}(t_k)]^{\mathrm{T}}$, $\bar{z}_i(t_k)$ is the initial nominal error state and $\bar{z}_{i,i-1}(t_k) = z_{i-1}^a(t_k; t_k)$ is the initial nominal error state of ASV $i - 1$, $\bar{\mathbf{z}}_i(s; t_k) = [\bar{z}_i(s; t_k)\ \bar{z}_{i,i-1}(s; t_k)]^{\mathrm{T}}$ and $\bar{\boldsymbol{u}}_i = [\bar{u}_i(s; t_k)\ \bar{u}_{i,i-1}(s; t_k)]^{\mathrm{T}}$. The terminal cost function is defined as $F_i\left(\bar{\mathbf{z}}_i(t_k + T; t_k)\right) = \|\bar{z}_i(t_k + T; t_k)\|_{P_i} + \|\bar{z}_{i,i-1}(t_k + T; t_k)\|_{P_{i,i-1}}$ and the stage cost is defined as

$$\begin{aligned} & L_i\left(\bar{\mathbf{z}}_i(s; t_k), z_{i-1}^a(s; t_k), \bar{\boldsymbol{u}}_i(s; t_k)\right) \\ = & \|\bar{z}_i(s; t_k)\|_{Q_i} + \|\bar{z}_i(s; t_k) - z_{i-1}^a(s; t_k)\|_{H_i} + \|\bar{u}_i(s; t_k)\|_{R_i} + \|\bar{z}_i(s; t_k) - z_i^a(s; t_k)\|_{F_i} \\ & + \|\bar{z}_{i,i-1}(s; t_k)\|_{Q_{i,i-1}} + \|\bar{u}_{i,i-1}(s; t_k)\|_{R_{i,i-1}}, \end{aligned} \tag{7.10}$$

with weighting matrices $P_i \succeq 0$, $P_{i,i-1} \succeq 0$, $Q_i \succeq 0$, $Q_{i,i-1} \succeq 0$, $R_{i,i-1} \succ 0$, $H_i \succeq 0$, $F_i \succeq 0$, $R_i \succ 0$, $H_{-1} = H_0 = Q_{0,-1} = R_{0,-1} = 0$.

Remark 7.1 The cost term $\|\bar{z}_i(s; t_k) - z_{i-1}^a(s; t_k)\|_{H_i}$ in (7.10) aims to minimize the predecessor–follower spacing error (i.e., $x_i - x_{i-1} + d^x = z_i - z_{i-1}$), and also maintains the same velocity between itself and its preceding ASV $i - 1$. The cost term $\|\bar{z}_i(s; t_k) - z_i^a(s; t_k)\|_{F_i}$ in (7.10) is introduced for the purpose of suppressing the deviation between the nominal error state $\bar{z}_i(s; t_k)$ and the assumed error state $z_i^a(s; t_k)$. The cost term $\|\bar{z}_{i,i-1}(s; t_k)\|_{Q_{i,i-1}}$ is designed to approximate the error state trajectory of ASV $i - 1$.

For the nominal error system in (7.6), the following assumption of the terminal set is given:

Assumption 7.2 For the nominal error system of ASV i in (7.6), there exist an admissible terminal control law $\kappa_i(\cdot) \in \bar{\mathbb{U}}_i$ and a terminal invariant set $\Omega_i(\epsilon_i)$ with a constant $\epsilon_i > 0$, such that: 1) $\forall \bar{z}_i(t_k) \in \Omega_i(\epsilon_i)$, and by implementing $\kappa_i(\cdot)$, $\bar{z}_i(t) \in \Omega_i(\epsilon_i)$ with $t > t_k$; 2) $\forall \bar{z}_i(t) \in \Omega_i(\epsilon_i)$, $i \in \mathcal{V}^0$, $\sum_{i \in \mathcal{V}^0} \left\{\dot{F}_i\left(\bar{\mathbf{z}}_i(t)\right) + L_i\left(\bar{\mathbf{z}}_i(t), z_{i-1}^a(t), \bar{\boldsymbol{u}}_i(t)\right)\right\} \leq 0$.

7.3 Robust Distributed Model Predictive Platooning Control

In this section, the DMPC optimization problem and the auxiliary control law design are firstly presented. The robust control invariant set is then formulated in Lemma 1. Finally, the robust DMPC algorithm is presented.

7.3.1 DMPC Optimization Problem

For ASV i, $i \in \mathcal{V}^0$ at time t_k, the DMPC optimization problem $\mathcal{P}_i$ is designed as

$$\min_{\bar{\boldsymbol{u}}_i(\cdot;t_k)} J_i\left(\bar{z}_i(t_k), z^a_{i-1}(\cdot;t_k), \bar{\boldsymbol{u}}_i(\cdot;t_k)\right)$$

$$\text{s.t. for } s \in [t_k, t_k+T),$$

$$\bar{z}_i(t_k;t_k) = \bar{z}_i(t_k), \tag{7.11a}$$

$$\dot{\bar{z}}_i(s;t_k) = f_i(\bar{z}_i(s;t_k)) + B_i\bar{u}_i(s;t_k), \tag{7.11b}$$

$$\bar{u}_i(s;t_k) \in \bar{\mathbb{U}}_i, \tag{7.11c}$$

$$\|\bar{z}_i(s;t_k) - z^a_i(s;t_k)\| \le \check{o}, \tag{7.11d}$$

$$\bar{z}_i(t_k+T;t_k) \in \Omega_i(\epsilon_i), \tag{7.11e}$$

$$\bar{z}_{i,i-1}(t_k;t_k) = z^a_{i-1}(t_k;t_k), \tag{7.11f}$$

$$\dot{\bar{z}}_{i,i-1}(s;t_k) = f_{i-1}(\bar{z}_{i,i-1}(s;t_k)) + B_{i-1}\bar{u}_{i,i-1}(s;t_k), \tag{7.11g}$$

$$\bar{u}_{i,i-1}(s;t_k) = u^a_{i-1}(s;t_k),\ s \in [t_k, t_{k+1}), \tag{7.11h}$$

$$\bar{u}_{i,i-1}(s;t_k) \in \bar{\mathbb{U}}_{i-1}, \tag{7.11i}$$

$$\bar{z}_{i,i-1}(t_k+T;t_k) \in \Omega_{i-1}(\epsilon_{i-1}), \tag{7.11j}$$

$$E(\bar{z}_i(s;t_k) - \bar{z}_{i,i-1}(s;t_k)) \le d^x - \hat{d}^{\text{safe}}_i, \tag{7.11k}$$

where the tightened control input constraint $\bar{\mathbb{U}}_i$ is defined as $\bar{\mathbb{U}}_i = \mathbb{U}_i \ominus K_i\mathbb{E}_i \neq \emptyset$, where K_i is the control gain of the auxiliary controller and $\mathbb{E}_i$ is the robust control invariant set, and $\hat{d}^{\text{safe}}_i := d^{\text{safe}}_i + e^x_{i,\max} + e^x_{i-1,\max} + \check{o}$. $e^x_{i,\max}$ is the maximum position error between the nominal position and the actual position, which will be given in Sect. 7.3.2. Based on the assumed state $z^a_{i-1}(t_k;t_k)$ transmitted from ASV $i-1$, the state trajectory of ASV $i-1$ is locally estimated by ASV i. The terminal constraint is defined as $\Omega_i(\epsilon_i) = \{\bar{z}_i | \|\bar{z}_i\| \le \epsilon_i\}$, where ϵ_i also satisfies $\epsilon_i < \check{o}$, and $\epsilon_i + \epsilon_{i-1} < d^x - \hat{d}^{\text{safe}}_i$.

Remark 7.2 The intervehicle collision avoidance safety of cooperative autonomous marine vehicles in Wei et al. (2021) is considered by the design of a collision avoidance cost term. In contrast, the intervehicle safety of the ASV platoon is guaranteed via a coupled constraint in (7.11k). The presence of coupled constraints makes the

DMPC optimization problem more challenging and interesting in terms of the recursive feasibility analysis. Following the similar approximation strategy in Richards and How (2007); Trodden and Richards (2013), the nominal error dynamics and constraints of ASV $i-1$ are assumed to be known by the local ASV i in optimization problem $\mathcal{P}_i$. Hence, ASV i is able to approximate the error state trajectory of its preceding ASV $i-1$, which will not be transmitted among ASVs. In this way, the coupled constraint (7.8) can be decoupled as a local constraint (7.11k).

Remark 7.3 The assumed state trajectory $z_{i-1}^a(\cdot; t_k)$ from ASV $i-1$ to ASV i is involved in two aspects: (1) The design of the predecessor–follower spacing error cost term in (7.10); (2) the initial value in (7.11f). Because the assumed state trajectory z_i^a in (7.9) is simultaneously transmitted among ASVs, the nominal control input for each ASV can be determined by solving the local optimization problem in parallel.

7.3.2 *The Auxiliary Controller*

The control input that will be applied to ASV i is designed as

$$u_i(s) = \bar{u}_i^*(s; t_k) + \pi_i(z_i(s), \bar{z}_i^*(s; t_k)), s \in [t_k, t_{k+1}). \tag{7.12}$$

For the sake of simplicity, $\bar{u}_i^*(s)$ is used to denote the optimal nominal control $\bar{u}_i^*(s; t_k)$ and $\pi_i(s)$ is used to the auxiliary control law $\pi_i(z_i(s), \bar{z}_i^*(s; t_k))$ in the sequel. The deviation between the actual and nominal error system state can be expressed as

$$e_i(s) = z_i(s) - \bar{z}_i(s; t_k), s \geq t_k, \tag{7.13}$$

where $e_i(s) = \left[e_{xi}(s)\; e_{\mu i}(s)\right]^{\mathrm{T}}$.

In the next, the robust control invariant set $\mathbb{E}_i$ is derived. The set $\mathbb{E}_i$ centers along the reference state trajectory (i.e., the nominal error state trajectory $\bar{z}_i(s; t_k)$) and the auxiliary control law $\pi_i(s)$ ensures that $e_i(s) \in \mathbb{E}_i$.

The following lemma gives the design of the auxiliary controller and the robust control invariant set $\mathbb{E}_i$:

Lemma 7.1 *Suppose that Assumption 7.1 holds. Let* $S_i(e_i(s)) : \mathbb{R}^2 \to [0, \infty)$ *be a continuously differentiable function with* $0 < S_i(e_i(s)) \leq \alpha_2(\|e_i(s)\|)$, *where* $\alpha_2(\cdot)$ *is a* $\mathcal{K}_\infty$ *function. Let* $\bar{z}_i(s; t_k)$, $s \geq t_k$ *be the reference state trajectory. The feedback control law is defined as*

$$\pi_i(s) := -K_i e_i(s), \tag{7.14}$$

with $K_i = [k_{pi} + k_{di} + d_i,\; M_{x_i} + k_{di}]$, $d_i = -X_{\mu_i} - X_{\mu_i|\mu_i|}|\mu_i|$. *If there exist* K_i, $\rho_i > 0$, *and* $\sigma_i > 0$, *such that*

$$\frac{\mathrm{d}}{\mathrm{d}s} S_i(e_i(s)) + \rho_i S_i(e_i(s)) - \sigma_i w_i^2(s) \leq 0, \forall w_i(s) \in \mathbb{W}_i, \tag{7.15}$$

where $S_i(e_i(s)) = \frac{1}{2}e_i^T(s)\Gamma_i e_i(s)$, $\Gamma_i = [k_{pi}, M_{x_i}; M_{x_i}, M_{x_i}]$. *Then,* $\forall e_i(t_k) \in \mathbb{E}_i$, $e_i(s) \in \mathbb{E}_i$, *for all* $s > t_k$, *where*

$$\mathbb{E}_i := \left\{ e_i(s) | S_i(e_i(s)) \leq \frac{\sigma_i w_{i,\max}^2}{\rho_i} \right\}. \tag{7.16}$$

Proof The proof can be obtained by employing the argument in Shuyou et al. (2013) [Lemma 1]. Also, it implies that the deviations $|e_i^x(s)|$ and $|e_i^\mu(s)|$ are bounded, i.e., $|e_i^x(s)| \leq e_{i,\max}^x$, $|e_i^\mu(s)| \leq e_{i,\max}^\mu$, with $e_{i,\max}^x > 0$ and $e_{i,\max}^\mu > 0$. □

The robust DMPC algorithm is applied in a receding horizon fashion and is presented in Algorithm 7.1.

Algorithm 7.1 Robust DMPC Algorithm

1: **Initialization**: Set $k = 0$. For ASV i, $i \in \mathcal{V}^0$, give the desired state trajectory $q_i^{\text{des}}(\cdot; t_k)$ at t_k. Set $\bar{z}_0(t_k) = 0$ and $\bar{z}_i(s; t_k) = z_i^a(s; t_k) = 0$, $s \in [t_k, t_k + T]$;
2: ASV 0 samples $z_0(t_k)$, solves $\mathcal{P}_0$ without (7.11f)–(7.11k) and broadcasts $z_0^a(\cdot; t_k)$ to ASV j, $j \in \mathcal{V}$;
3: ASV i, $i \in \mathcal{V}$ receives $z_\ell^a(\cdot; t_k)$, $\ell \in \mathcal{N}_i$;
4: ASV i, $i \in \mathcal{V}$ samples $z_i(t_k)$, solves $\mathcal{P}_i$ in (7.11), and further generates control input trajectory $u_i(s)$ via (7.12);
5: ASV i, $i \in \mathcal{V}$ applies $u_i(s)$, $s \in [t_k, t_k + \delta)$, and transmits $z_i^a(\cdot; t_k)$ to ASV $i + 1$.
6: $k = k + 1$. Go to Step 2.

7.4 Theoretical Analysis

In the following, the theoretical ana lysis is presented, including the recursive feasibility of the proposed algorithm and the closed-loop robust stability of the heterogeneous ASV platoon with external disturbances.

Theorem 7.1 *Suppose that Assumptions 7.1 and 7.2 hold. For the heterogeneous ASV platoon, if the optimization problem* $\mathcal{P}_i$ *in (7.11) is feasible at* t_0, *then,*
(1) For ASV i, $i \in \mathcal{V}^0$, *problem* $\mathcal{P}_i$ *is recursively feasible at time* t_k, $k \geq 1$ *by application of Algorithm 7.1 if the following conditions hold:* $\epsilon_i < \check{o}$, *and* $\epsilon_i + \epsilon_{i-1} < d^x - \hat{d}_i^{safe}$.
(2) Furthermore, the actual intervehicle safety can be ensured for all admissible disturbances $w_i \in \mathbb{W}_i$.
(3) The overall closed-loop system with the robust DMPC strategy is ISS if the following condition holds: $F_i \succeq H_{i+1}$.

Proof (1) The recursive feasibility of the DMPC algorithm is proved by induction. The optimal nominal control input trajectories obtained at t_k are denoted as $\bar{u}_i^*(s; t_k)$, $\bar{u}_{i,i-1}^*(s; t_k)$, $s \in [t_k, t_k + T)$, which can drive the nominal error systems of

ASV i and ASV $i-1$ into the terminal region $\Omega_i(\epsilon_i)$ and $\Omega_{i-1}(\epsilon_{i-1})$, respectively. The optimal nominal error state trajectory with control input trajectory $\bar{u}_i^*(s;t_k)$, $s \in [t_k, t_k+T)$ is given by $\dot{\bar{z}}_i^*(s;t_k) = f_i(\bar{z}_i^*(s;t_k)) + B_i\bar{u}_i^*(s;t_k)$, $s \in [t_k, t_k+T)$. At t_{k+1}, based on the optimal nominal control input trajectory at t_k and Assumption 7.2, a feasible control input trajectory candidate $\bar{u}_i^p(\cdot;t_{k+1})$ for the problem $\mathcal{P}_i$ is constructed as

$$\bar{u}_i^p(s;t_{k+1}) = \begin{cases} \bar{u}_i^*(s;t_k), & s \in \left[t_{k+1}, t_k+T\right), \\ \kappa_i(\bar{z}_i(s;t_k)), & s \in \left[t_k+T, t_{k+1}+T\right), \end{cases} \tag{7.17}$$

where $\kappa_i(\cdot)$ is the terminal controller for ASV i. The corresponding state trajectory is computed as

$$\bar{z}_i^p(s;t_{k+1}) = \begin{cases} \bar{z}_i^*(s;t_k), & s \in \left[t_{k+1}, t_k+T\right), \\ z_i^\kappa(s;t_k), & s \in \left[t_k+T, t_{k+1}+T\right). \end{cases} \tag{7.18}$$

The control input trajectory candidate and the corresponding state trajectory of locally approximated ASV $i-1$ are constructed in a similar way.

According to Assumption 7.2, $\kappa_i(\bar{z}_i(s;t_k)) \in \bar{\mathbb{U}}_i$, $s \in [t_k+T, t_{k+1}+T)$. In addition, $\bar{u}_i^p(s;t_{k+1}) = \bar{u}_i^*(s;t_k) \in \bar{\mathbb{U}}_i$, $s \in [t_{k+1}, t_k+T)$. Hence, the control input constraint (7.11c) satisfaction is obvious. Further, since the terminal region $\Omega_i(\epsilon_i)$ is invariant with the terminal control law $\kappa_i(\bar{z}_i(s;t_k))$, $s \in [t_k+T, t_{k+1}+T)$, it has $\bar{z}_i^p(t_{k+1}+T;t_{k+1}) \in \Omega_i(\epsilon_i)$. The terminal state constraint (7.11e) is satisfied. Due to the assumed state trajectory in (7.9) and the state trajectory in (7.18) coincide, i.e., $\bar{z}_i^p(s;t_{k+1}) = z_i^a(s;t_{k+1})$, $s \in [t_{k+1}, t_k+T]$. In addition, $z_i^a(s;t_k) = 0$, $s \in (t_k+T, t_{k+1}+T]$ and $\epsilon_i \le \check{o}$, thus $\|\bar{z}_i^p(s;t_{k+1}) - z_i^a(s;t_{k+1})\| \le \check{o}$, $s \in (t_k+T, t_{k+1}+T]$ holds. Hence, the constraint (7.11d) is fulfilled. The constraints (7.11i) and (7.11j) of ASV $i-1$ can be proved following the same logic of (7.11c) and (7.11e). Next, for $s \in [t_{k+1}, t_k+T]$,

$$\begin{aligned} & E(\bar{z}_i^p(s;t_{k+1}) - \bar{z}_{i,i-1}^p(s;t_{k+1})) \\ =& E(\bar{z}_i^*(s;t_k) - \bar{z}_{i,i-1}^*(s;t_k)) \\ \le& d^x - \hat{d}_i^{\text{safe}}, \end{aligned}$$

and for $s \in (t_k+T, t_{k+1}+T]$, since the inequality $\epsilon_i + \epsilon_{i-1} < d^x - \hat{d}_i^{\text{safe}}$ holds, and the terminal constraint implies that (7.11k) is ensured. Thus, the minimum safe following distance constraint (7.11k) is satisfied.

(2) At time t_k, from Lemma 7.1, it can be obtained $z_i(s) \in \bar{z}_i^*(s;t_k) \oplus \mathbb{E}_i$, $s \in [t_k, t_k+\delta]$, $z_{i-1}(s) \in \bar{z}_{i-1}^*(s;t_k) \oplus \mathbb{E}_{i-1}$, $s \in [t_k, t_k+\delta]$. Since $z_{i-1}^a(t_k;t_k) = \bar{z}_{i,i-1}(t_k;t_k) = \bar{z}_{i-1}(t_k;t_k)$, and $\bar{u}_{i,i-1}^*(s;t_k) = u_{i-1}^a(s;t_k)$, then $\bar{z}_{i,i-1}(s;t_k) = z_{i-1}^a(s;t_k)$, $s \in [t_k, t_k+\delta]$. In addition, by the constraint (7.11d), it can be obtained $z_{i-1}(s) \in \bar{z}_{i,i-1}^*(s;t_k) \oplus \mathbb{E}_{i-1} \oplus \check{\mathbb{O}}$, $s \in [t_k, t_k+\delta]$, $\check{\mathbb{O}} := \{\bar{z}_i \| \|\bar{z}_i - z_i^a\| \le \check{o}\}$. Therefore, the actual inter-ASV safety in (7.8) is ensured since the constraint (7.11k) holds.

(3) For the overall ASV system, the Lyapunov function is chosen as the sum of the distributed optimal cost function

$$V(t_k) = \sum_{i \in \mathcal{V}^0} J_i\left(\bar{z}_i(t_k), z_{i-1}^a(\cdot; t_k), \bar{\boldsymbol{u}}_i^*(\cdot; t_k)\right). \tag{7.19}$$

At time t_k, the calculated optimal control input trajectory $\bar{u}_i^*(s; t_k)$,$s \in [t_k, t_k + \delta)$ is applied. At t_{k+1}, applying a suboptimal feasible control input trajectory (7.17) yields

$$\begin{aligned}
V(t_{k+1}) &= \sum_{i \in \mathcal{V}^0} J_i\left(\bar{\mathbf{z}}_i(t_{k+1}), z_{i-1}^a(\cdot; t_{k+1}), \bar{\mathbf{u}}_i^*(\cdot; t_{k+1})\right) \\
&\le \sum_{i \in \mathcal{V}^0} J_i\left(\bar{\mathbf{z}}_i(t_{k+1}), z_{i-1}^a(\cdot; t_{k+1}), \bar{\mathbf{u}}_i^p(\cdot; t_{k+1})\right) \\
&= \sum_{i \in \mathcal{V}^0} \left\{ \int_{t_{k+1}}^{t_{k+1}+T} L_i(\bar{\mathbf{z}}_i^p(s; t_{k+1}), z_{i-1}^a(s; t_{k+1}), \bar{\mathbf{u}}_i^p(s; t_{k+1}))\mathrm{d}s + F_i(\bar{\mathbf{z}}_i^p(t_{k+1}+T; t_{k+1})) \right\}.
\end{aligned}$$

Hence, the difference of the Lyapunov function (7.19) at t_{k+1} and t_k becomes

$$\begin{aligned}
&V(t_{k+1}) - V(t_k) \\
&\le \sum_{i \in \mathcal{V}^0} J_i\left(\bar{z}_i(t_{k+1}), z_{i-1}^a(\cdot; t_{k+1}), \bar{\boldsymbol{u}}_i^p(\cdot; t_{k+1})\right) - V(t_k) \\
&= \sum_{i \in \mathcal{V}^0} (\Delta_1^i + \Delta_2^i + \Delta_3^i),
\end{aligned}$$

where

$$\begin{aligned}
\Delta_1^i =& -\int_{t_k}^{t_{k+1}} \{\|\bar{z}_i^*(s; t_k) - z_{i-1}^a(s; t_k)\|_{H_i} + \|\bar{z}_i^*(s; t_k) - z_i^a(s; t_k)\|_{F_i} \\
&+ \|\bar{z}_i^*(s; t_k)\|_{Q_i} + \|\bar{u}_i^*(s; t_k)\|_{R_i} + \|\bar{z}_{i,i-1}^*(s; t_k)\|_{Q_{i,i-1}} + \|\bar{u}_{i,i-1}^*(s; t_k)\|_{R_{i,i-1}}\}\mathrm{d}s, \\
\Delta_2^i =& \int_{t_{k+1}}^{t_k+T} \{L_i(\bar{\mathbf{z}}_i^p(s; t_{k+1}), z_{i-1}^a(s; t_{k+1}), \bar{\mathbf{u}}_i^p(s; t_{k+1})) - L_i(\bar{\mathbf{z}}_i^*(s; t_k), z_{i-1}^a(s; t_k), \bar{\mathbf{u}}_i^*(s; t_k))\}\mathrm{d}s, \\
\Delta_3^i =& \int_{t_k+T}^{t_{k+1}+T} \{\|\bar{z}_i^p(s; t_{k+1})\|_{Q_i} + \|\bar{u}_i^p(s; t_{k+1})\|_{R_i} + \|\bar{u}_{i,i-1}^p(s; t_{k+1})\|_{R_{i,i-1}} + \|\bar{z}_{i,i-1}^p(s; t_{k+1})\|_{Q_{i,i-1}} \\
&+ \|\bar{z}_i^p(s; t_{k+1})\|_{F_i} + \|\bar{z}_i^p(s; t_{k+1})\|_{H_i}\}\mathrm{d}s + \|\bar{z}_i^p(t_{k+1}+T; t_{k+1})\|_{P_i} - \|\bar{z}_{i,i-1}^*(t_k+T; t_k)\|_{P_{i,i-1}} \\
&+ \|\bar{z}_{i,i-1}^p(t_{k+1}+T; t_{k+1})\|_{P_{i,i-1}} - \|\bar{z}_i^*(t_k+T; t_k)\|_{P_i}.
\end{aligned}$$

It is easy to find that $\Delta_1^i < 0$ always holds. Since the assumed state trajectory in (7.9) and the state trajectory in (7.18) coincide, that is, $\bar{z}_i^p(s; t_{k+1}) = z_i^a(s; t_{k+1}), s \in [t_{k+1}, t_k + T]$. Then, it can be obtained

$$
\begin{aligned}
&\|\bar{z}_i^p(t_{k+1}^s) - z_i^a(t_{k+1}^s)\|_{F_i} - \|\bar{z}_i^*(t_k^s) - z_{i-1}^a(t_k^s)\|_{H_i} + \|\bar{z}_i^p(t_{k+1}^s) - z_{i-1}^a(t_{k+1}^s)\|_{H_i} - \|\bar{z}_i^*(t_k^s) - z_i^a(t_k^s)\|_{F_i} \\
=&\|\bar{z}_i^p(t_{k+1}^s) - z_{i-1}^a(t_{k+1}^s)\|_{H_i} - \|\bar{z}_i^*(t_k^s) - z_i^a(t_k^s)\|_{F_i} - \|\bar{z}_i^*(t_k^s) - z_{i-1}^a(t_k^s)\|_{H_i} \\
\le&\|\bar{z}_i^*(t_k^s) - z_{i-1}^a(t_k^s)\|_{H_i} + \|\bar{z}_{i-1}^*(t_k^s) - z_{i-1}^a(t_k^s)\|_{H_i} - \|\bar{z}_i^*(t_k^s) - z_i^a(t_k^s)\|_{F_i} - \|\bar{z}_i^*(t_k^s) - z_{i-1}^a(t_k^s)\|_{H_i} \\
=&\|\bar{z}_{i-1}^*(t_k^s) - z_{i-1}^a(t_k^s)\|_{H_i} - \|\bar{z}_i^*(t_k^s) - z_i^a(t_k^s)\|_{F_i},
\end{aligned} \tag{7.20}
$$

where (t_{k+1}^s) is used to denote $(s; t_{k+1})$.

Since $H_0 = H_{-1} = 0$ and $F_i \succeq H_{i+1}$, for Δ_2^i,

$$
\begin{aligned}
\sum_{i\in\mathcal{V}^0} \Delta_2^i =& \sum_{i\in\mathcal{V}^0} \left\{ \int_{t_{k+1}}^{t_k+T} \{ - \|\bar{z}_i^*(s; t_k) - z_i^a(s; t_k)\|_{F_i} + \|\bar{z}_i^p(s; t_{k+1}) - z_{i-1}^a(s; t_{k+1})\|_{H_i} \right. \\
& \left. - \|\bar{z}_i^*(s; t_k) - z_{i-1}^a(s; t_k)\|_{H_i} \}\mathrm{d}s \right\} \\
\overset{(7.20)}{\le}& \sum_{i\in\mathcal{V}^0} \left\{ \int_{t_{k+1}}^{t_k+T} \{\|\bar{z}_{i-1}^*(s; t_k) - z_{i-1}^a(s; t_k)\|_{H_i} - \|\bar{z}_i^*(s; t_k) - z_i^a(s; t_k)\|_{F_i} \}\mathrm{d}s \right\} \\
\le& \int_{t_{k+1}}^{t_k+T} -\|\bar{z}_N^*(s; t_k) - z_N^a(s; t_k)\|_{F_N} \mathrm{d}s \le 0.
\end{aligned} \tag{7.21}
$$

With the terminal control law $\kappa_i(\cdot)$ and $\kappa_{i-1}(\cdot)$, the system states will be $\bar{z}_i^p(s; t_{k+1})$ and $\bar{z}_{i,i-1}^p(s; t_{k+1})$, $s \in [t_k + T, t_{k+1} + T]$, respectively. By Assumption 7.2, it follows that

$$
\begin{aligned}
&\sum_{i\in\mathcal{V}^0} \Delta_3^i \\
=& \sum_{i\in\mathcal{V}^0} \{ \int_{t_k+T}^{t_{k+1}+T} \{\|\bar{z}_i^p(s; t_{k+1})\|_{Q_i} + \|\bar{u}_i^p(s; t_{k+1})\|_{R_i} + \|\bar{z}_i^p(s; t_{k+1})\|_{F_i} + \|\bar{z}_{i,i-1}^p(s; t_{k+1})\|_{Q_{i,i-1}} \\
& + \|\bar{z}_i^p(s; t_{k+1})\|_{H_i} + \|\bar{u}_{i,i-1}^p(s; t_{k+1})\|_{R_{i,i-1}} \}\mathrm{d}s \\
& + \|\bar{z}_i^p(t_{k+1} + T; t_{k+1})\|_{P_i} - \|\bar{z}_{i,i-1}^*(t_k + T; t_k)\|_{P_{i,i-1}} \\
& + \|\bar{z}_{i,i-1}^p(t_{k+1} + T; t_{k+1})\|_{P_{i,i-1}} - \|\bar{z}_i^*(t_k + T; t_k)\|_{P_i} \} \\
\le& 0.
\end{aligned} \tag{7.22}
$$

According to (7.20), (7.21) and (7.22), it can be obtained

$$
V(t_{k+1}) - V(t_k) \le \sum_{i\in\mathcal{V}^0} (\Delta_1^i + \Delta_2^i + \Delta_3^i) < 0, \tag{7.23}
$$

which means that the nominal error systems are asymptotically stable. Define $\bar{Z}^*(t) = [\bar{z}_0^{*\mathrm{T}}(t), \ldots, \bar{z}_N^{*\mathrm{T}}(t)]^\mathrm{T}$, $Z(t) = [z_0^\mathrm{T}(t), z_1^\mathrm{T}(t), \ldots, z_N^\mathrm{T}(t)]^\mathrm{T}$, $E(t) = [e_0^\mathrm{T}(t), e_1^\mathrm{T}(t), \ldots, e_N^\mathrm{T}(t)]^\mathrm{T}$, $U(t) = [u_0(t), \ldots, u_N(t)]^\mathrm{T}$ and $\bar{U}^*(t) = [\bar{u}_0^*(t), \ldots, \bar{u}_N^*(t)]^\mathrm{T}$, $W(t) = [w_0(t), \ldots, w_N(t)]^\mathrm{T}$. Further, the overall closed-loop nominal system and the overall closed-loop actual system are given by

$$\dot{\bar{Z}}^*(t) = F(\bar{Z}^*(t)) + B\bar{U}^*(t) \tag{7.24a}$$

$$\dot{Z}(t) = F(Z(t)) + BU(t) + BW(t), \tag{7.24b}$$

where $F(\cdot) = [(f_0(z_0(t)))^{\mathrm{T}}, \ldots, (f_N(z_N(t)))^{\mathrm{T}}]^{\mathrm{T}} \in \mathbb{R}^{2(N+1)\times 1}$ and $B = \text{diag}(B_0, \ldots, B_N) \in \mathbb{R}^{2(N+1)\times N}$. By Khalil (1996) [Lemma 4.5], if the overall nominal error system in (7.24a) is asymptotically stable, then there exists a $\mathcal{KL}$ function $\alpha_3(\cdot, \cdot)$, such that

$$\|\bar{Z}^*(t)\| \leq \alpha_3(\|\bar{Z}^*(t_0)\|, t - t_0), t \geq t_0. \tag{7.25}$$

By Lemma 7.1, if $e_i(t_k) \in \mathbb{E}_i$, then $e_i(t) \in \mathbb{E}_i$, $t \geq t_k$ always holds. The inequality $\frac{1}{2}e_i(t)^{\mathrm{T}}\Gamma_i e_i(t) \leq \frac{\sigma_i w_{i,\max}^2}{\rho_i}$ results in $\|e_i(t)\| \leq c_i w_{i,\max}$, where $c_i = \sqrt{\frac{2\sigma_i}{\rho_i \lambda_{\min}(\Gamma_i)}}$. By defining a $\mathcal{K}$ function $\gamma_i(\|w_i\|_{[t_0,t]}) = c_i w_{i,\max}$, it can be obtained $\|e_i(t)\| \leq \gamma_i(\|w_i\|_{[t_0,t]})$. By noting that if the actual error state $z_i(t)$ satisfies $z_i(t) \in \bar{z}_i(t) \oplus \mathbb{E}_i$ for all admissible disturbance sequences. It implies that $\|z_i(t)\| \leq \|\bar{z}_i(t)\| + c_i w_{i,\max}$ holds. Since $z_i(t) = \bar{z}_i^*(t) + e_i(t)$, for the overall platoon system it can be obtained that $Z(t) = \bar{Z}^*(t) + E(t)$. By defining a $\mathcal{K}$ function $\gamma(\|W\|_{[t_0,t]}) = c\sqrt{N+1}\|W_{\max}\|$ for the overall closed-loop system, it is easy to obtain $\|E(t)\| \leq \sqrt{\sum_{i\in\mathcal{V}^0} c_i^2 w_{i,\max}^2} \leq \gamma(\|W\|_{[t_0,t]})$, where $c = \max_{i\in\mathcal{V}^0}\{c_i\}$ and $W_{\max} = [w_{0,\max}, \ldots, w_{N,\max}]^{\mathrm{T}}$. Therefore, it follows from (7.25) that

$$\|Z(t)\| \leq \|\bar{Z}^*(t)\| + \|E(t)\| \overset{(7.25)}{\leq} \alpha_3(\|\bar{Z}^*(t_0)\|, t - t_0) + \gamma(\|W\|_{[t_0,t]}). \tag{7.26}$$

According to Definition 7.1, the overall closed-loop system in (7.24b) is proved to be ISS. □

The stability analysis is inspired by Dunbar and Caveney (2012); Wang and Ding (2014), which gives the stability proof for the closed-loop system without external disturbances. In this work, the closed-loop stability is proved based on the sum of the distributed cost function at two consecutive time instants t_k and t_{k+1}. However, the stability during the period $[t_k, t_{k+1}]$, $k \in \mathbb{N}$ should also be guaranteed. This will be discussed in future research.

7.5 Simulation Study

In this section, the simulation of the heterogeneous ASV platoon is presented to verify the effectiveness of the proposed robust distributed model predictive platooning control strategy. The simulated multi-ASV platoon consists of seven Cybership II with different configuration parameters (Skjetne et al. 2005), which are selected and shown in Table 7.1. The control input constraint is $|u_i| \leq 40$N. The desired spac-

Table 7.1 Main parameters of the ASVs in the platoon

ASV i	Mass m_i (kg)	Added mass $X_{\dot{\mu}_i}$ (kg)	Surge linear drag X_{μ_i} (kg/s)	Surge quadratic drag $X_{\mu_i\|\mu_i\|}$ (kg/m)
0	15.0	−2.00	−0.72	−2.05
1	18.0	−2.00	−0.80	−2.00
2	12.0	−3.00	−1.30	−1.60
3	14.0	−1.60	−1.70	−2.30
4	13.7	−2.60	−1.35	−1.80
5	13.0	−1.30	−1.00	−1.30
6	14.0	−3.60	−1.20	−1.30

Table 7.2 Disturbance parameters of the ASV platoon

i	0	1	2	3	4	5	6
A_i	4.5	5.0	5.5	5.3	6.1	5.89	6.0
b_i	0.5π	0.7π	0.8π	0.75π	0.9π	0.8π	0.65π
c_i	$\pi/2$	0	$\pi/3$	$-\pi/4$	$\pi/5$	$2\pi/3$	$-\pi/6$

ing is chosen as $d^x = 2.5$m, and the minimum safe following distance is given as $d_i^{\text{safe}} = 0.5$m.

The desired speed $\mu_0(t)$ is given by

$$\mu_0(t) = \begin{cases} 1 \text{ m/s}, & 0\text{s} \leq t < 1.6\text{s}, \\ 1 + 1.4t \text{ m/s}, & 1.6\text{s} < t \leq 2\text{s}, \\ 1.56 \text{ m/s}, & 2\text{s} < t \leq 3.9\text{s}, \\ 1.56 - 1.6t \text{ m/s}, & 3.9\text{s} < t \leq 4.4\text{s}, \\ 0.76 \text{ m/s}, & 4.4\text{s} < t \leq 6\text{s}. \end{cases} \tag{7.27}$$

In the simulation, time-varying disturbances are simulated by $w_i(t) = A_i \sin(b_i t + c_i)$, where A_i, b_i, and c_i, $i \in \mathcal{V}^0$ are given in Table 7.2.

The parameters for the DMPC optimization problem $\mathcal{P}_i$ are designed as follows: The sampling period $\delta = 0.1$s, the prediction horizon $T = 5\delta$. The weighting matrices are chosen as, $Q_i = Q_{i,i-1} = \text{diag}(10^3, 10^2)$, $P_i = P_{i,i-1} = \text{diag}(10^4, 10^3)$, $H_i = \text{diag}(10^2, 10)$, and $F_i = \text{diag}(10^2, 10^2)$. Note that the weighting matrix $R_i = R_{i,i-1} = 10^{-4}$ is chosen so that the control penalty is comparable to the cost terms of Q_i, F_i, and H_i. The terminal set level is $\epsilon_i = 0.2$, and the upper bound of the predicted and assumed error state is chosen as $\breve{o} = 0.5$.

The robust control invariant set $\mathbb{E}_i$ in (7.16) is constructed based on the conditions in Lemma 7.1. In addition, the auxiliary controller should be designed to guarantee that the tightened control input constraint set is not empty, i.e., $\bar{\mathbb{U}}_i = \mathbb{U}_i \ominus K_i\mathbb{E}_i \neq \emptyset$.

Table 7.3 Auxiliary control gain K_i of the ASVs

ASV i	0	1	2	3	4	5	6
k_{pi}	6.0	7.0	9.0	4.0	9.0	10.0	8.0
k_{di}	6.0	8.0	6.0	4.0	8.0	8.0	7.0
ρ_i	0.92	0.85	1.32	1.13	0.75	1.0	0.76
σ_i	0.20	0.10	0.25	0.15	0.05	0.08	0.09

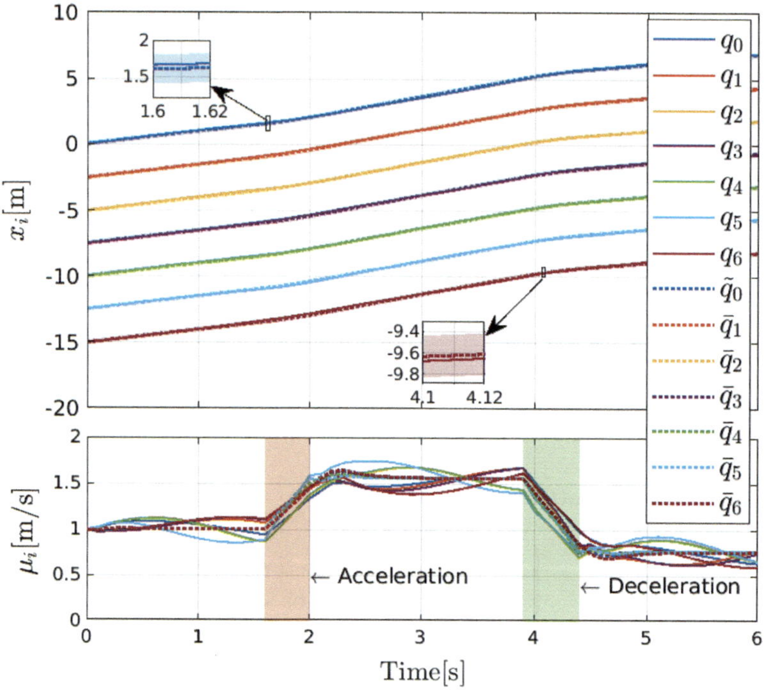

Fig. 7.1 State trajectory q_i and $\bar{q}_i$ of the ASVs by using the proposed robust DMPC

Intuitively, a smaller robust control invariant set leads to a larger admissible control input domain of the DMPC problem $\mathcal{P}_i$. The control gains k_{pi}, k_{di} and the variables σ_i, ρ_i for ASV $i, i \in \mathcal{V}^0$ are shown in Table 7.3. Then, the tightened control input constraint sets are: $|\bar{u}_0| \leq 20.67$N, $|\bar{u}_1| \leq 22.16$N, $|\bar{u}_2| \leq 21.32$N, $|\bar{u}_3| \leq 22.93$N, $|\bar{u}_4| \leq 26.24$N, $|\bar{u}_5| \leq 26.02$N, $|\bar{u}_6| \leq 21.38$N.

The simulation results of the heterogeneous ASV platoon using the proposed robust DMPC strategy are demonstrated in Figs. 7.1, 7.2 and 7.3. The resulting position trajectories of ASVs illustrate the good platooning control performance, as shown in Fig. 7.1. The dash line represents the nominal position of the ASV platoon, the solid line represents the actual position, and the shading area represents a hyper-tube whose center is the nominal state trajectory. As shown in Fig. 7.2, the error $e_i(t)$ between the nominal error state $\bar{z}_i(t)$ and the actual error state $z_i(t)$ stays within a

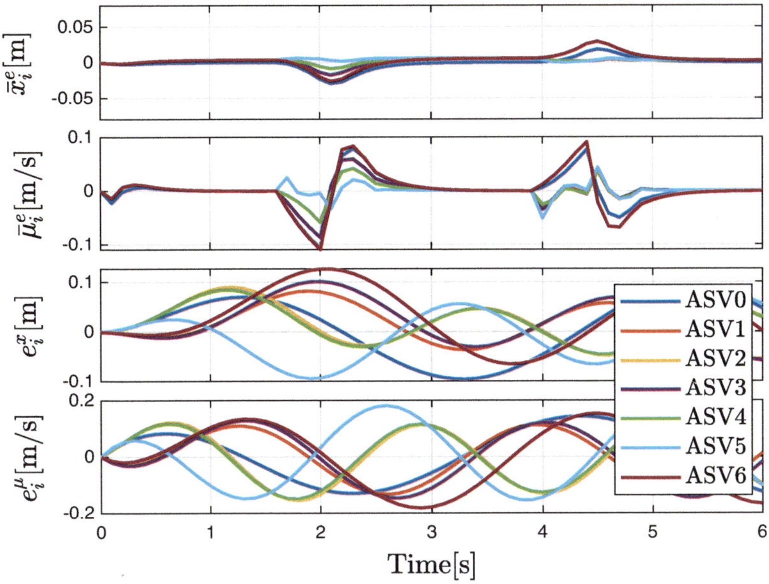

Fig. 7.2 Trajectories of $\bar{z}_i$ and e_i of the ASVs by using the proposed robust DMPC

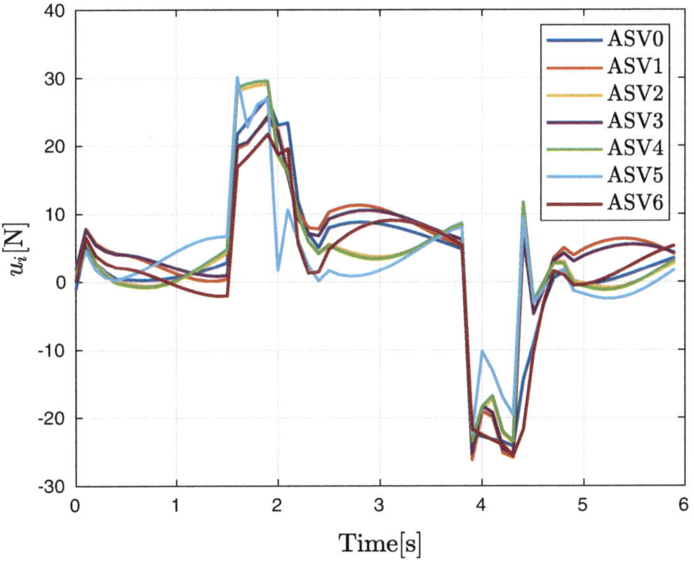

Fig. 7.3 Control input u_i of the ASVs by using the proposed robust DMPC

small bounded set and the nominal error state $\bar{z}_i(t)$ approximately converges to zero. Figure 7.3 shows the control input trajectories of seven heterogeneous ASVs. It can be observed that the control input constraint of each ASV can be always satisfied.

7.6 Notes and Summary

In this chapter, a robust DMPC approach is designed for the heterogeneous ASV platoon with external disturbances over the fixed predecessor–leader following communication topology. Each ASV receives the assumed state trajectory from its predecessor and the lead ASV, samples its states and broadcasts its predicted state trajectory to its following ASV. Further, by incorporating the coupled minimum safe following distance constraint into the DMPC optimization problem, the local safe control input is generated for each ASV. The recursive feasibility and closed-loop stability of the proposed method are rigorously analyzed. Simulation studies verify the effectiveness of the proposed distributed platooning control method. Future research can be focused on the design of resilient DMPC for the ASV platoon under cyber attacks, such as DoS attacks (Sun et al. 2019). Another interesting extension of this chapter is to investigate the ASV platoon with unknown model parameters using the adaptive MPC approach (Zhang and Shi 2020).

References

Alam A, Mårtensson J, Johansson KH (2015) Experimental evaluation of decentralized cooperative cruise control for heavy-duty vehicle platooning. Control Eng Pract 38:11–25

Besselink B, Johansson KH (2017) String stability and a delay-based spacing policy for vehicle platoons subject to disturbances. IEEE Trans Autom Control 62(9):4376–4391

Dunbar WB, Caveney DS (2012) Distributed receding horizon control of vehicle platoons: stability and string stability. IEEE Trans Autom Control 57(3):620–633

Feng S, Sun H, Zhang Y, Zheng J, Liu HX, Li L (2020) Tube-based discrete controller design for vehicle platoons subject to disturbances and saturation constraints. IEEE Trans Control Syst Technol 28(3):1066–1073

Feng S, Zhang Y, Li SE, Cao Z, Liu HX, Li L (2019) String stability for vehicular platoon control: definitions and analysis methods. Ann Rev Control 47:81–97

Fossen TI (2002) Marine control systems: guidance, navigation, and control of ships, rigs and underwater vehicles. Mar Cybern

Guanetti J, Kim Y, Borrelli F (2018) Control of connected and automated vehicles: state of the art and future challenges. Ann Rev Control 45:18–40

Jiang Z-P, Wang Y (2001) Input-to-state stability for discrete-time nonlinear systems. Automatica 37(6):857–869

Jin IG, Orosz G (2016) Optimal control of connected vehicle systems with communication delay and driver reaction time. IEEE Trans Intell Transp Syst 18(8):2056–2070

Khalil HK (1996) Nonlinear systems, 2nd edn. Prentice Hall, New York

Kianfar R, Falcone P, Fredriksson J (2015) A control matching model predictive control approach to string stable vehicle platooning. Control Eng Pract 45:163–173

Li H, Shi Y, Yan W (2016) Distributed receding horizon control of constrained nonlinear vehicle formations with guaranteed γ-gain stability. Automatica 68:148–154

Lin X, Görges D (2019) Robust model predictive control of linear systems with predictable disturbance with application to multiobjective adaptive cruise control. IEEE Trans Control Syst Technol 28(4):1460–1475

Liu P, Kurt A, Ozguner U (2018) Distributed model predictive control for cooperative and flexible vehicle platooning. IEEE Trans Control Syst Technol 27(3):1115–1128

Mayne DQ, Kerrigan EC, Van Wyk EJ, Falugi P (2011) Tube-based robust nonlinear model predictive control. Int J Robust Nonlinear Control 21(11):1341–1353

Mayne DQ, Seron MM, Raković SV (2005) Robust model predictive control of constrained linear systems with bounded disturbances. Automatica 41(2):219–224

Nikou A, Dimarogonas DV (2019) Decentralized tube-based model predictive control of uncertain nonlinear multiagent systems. Int J Robust Nonlinear Control 29(10):2799–2818

Richards A, How JP (2007) Robust distributed model predictive control. Int J Control 80(9):1517–1531

Skjetne R, Fossen TI, Kokotović PV (2005) Adaptive maneuvering, with experiments, for a model ship in a marine control laboratory. Automatica 41(2):289–298

Qi S, Zhang K, Shi Y (2019) Resilient model predictive control of cyber-physical systems under DoS attacks. IEEE Trans Ind Inform 16(7):4920–4927

Trodden P, Richards A (2010) Distributed model predictive control of linear systems with persistent disturbances. Int J Control 83(8):1653–1663

Trodden P, Richards A (2013) Cooperative distributed MPC of linear systems with coupled constraints. Automatica 49(2):479–487

Turri V, Besselink B, Johansson KH (2016) Cooperative look-ahead control for fuel-efficient and safe heavy-duty vehicle platooning. IEEE Trans Control Syst Technol 25(1):12–28

Viel C, Vautier U, Wan J, Jaulin L (2019) Platooning control for heterogeneous sailboats based on constant time headway. IEEE Trans Intell Transp Syst 21(5):2078–2089

Wang P, Ding B (2014) Distributed RHC for tracking and formation of nonholonomic multi-vehicle systems. IEEE Trans Autom Control 59(6):1439–1453

Wei H, Shen C, Shi Y (2021) Distributed Lyapunov-based model predictive formation tracking control for autonomous underwater vehicles subject to disturbances. IEEE Trans Syst Man Cybern Syst 51(8):5198–5208

Wu Y, Li SE, Cortés J, Poolla K (2020) Distributed sliding mode control for nonlinear heterogeneous platoon systems with positive definite topologies. IEEE Trans Control Syst Technol 28(4):1272–1283

Yu S, Maier C, Chen H, Allgöwer F (2013) Tube MPC scheme based on robust control invariant set with application to Lipschitz nonlinear systems. Syst Control Lett 62(2):194–200

Zhang K, Shi Y (2020) Adaptive model predictive control for a class of constrained linear systems with parametric uncertainties. Automatica 117:108974

Zheng Y, Li SE, Li K, Borrelli F, Hedrick JK (2016) Distributed model predictive control for heterogeneous vehicle platoons under unidirectional topologies. IEEE Trans Control Syst Technol 25(3):899–910

Chapter 8
Efficient Implementation Algorithms for NMPC-Based AUV Trajectory-Tracking Control

8.1 Introduction

8.1.1 Research Background and Contributions

In previous chapters, we have discussed the model predictive control for the motion control of a single AUV and for the cooperative control of a multi-vehicle system. The purpose of Chaps. 3–7 is to propose novel and useful frameworks that bring salient features, such as constraint handling and multiple control objective incorporation, in solving these AUV control problems. The main focus is hence on the design of the optimization problems such that the closed-loop system is theoretically stable with respect to the equilibrium of interest. The simulation results indicate that the MPC solutions are promising for the AUV applications due to their excellent control performance and robustness. However, there could be an issue that potentially prohibits the MPC from being implemented on AUV systems: The heavy computational burden.

As a feedback control technique, the MPC realizes the feedback mechanism by recursively solving open-loop OCPs with real-time state measurement as initial conditions. In theory, the solutions of the OCPs are assumed to be obtained instantly. In practice, they can only be provided by iterative methods. The OCPs need to be solved within strict time constraints. Violation of the time constraints possibly leads to performance degradation or even instability (Findeisen and Allgöwer 2004). This is exactly the reason why the most successful area for MPC is the chemical industry (Qin and Badgwell 1997) where the sampling period is sufficiently long for standard numerical algorithms. However, the AUV system is with fast dynamics, and the permitted sampling period may be less than a tenth of a second. Computation time is no longer negligible. Linear MPC, inheriting the good structure of the quadratic problem, might be efficiently solved by off-the-shelf numerical algorithms, while nonlinear MPC faces an intense conflict between the comparatively long computation time and the short sampling period. Due to the nonlinear dynamics of the AUV, the OCPs are essentially generic nonlinear programming (NLP) problems. The computational

Y. Shi et al., *Advanced Model Predictive Control for Autonomous Marine Vehicles*, Advances in Industrial Control, https://doi.org/10.1007/978-3-031-19354-5_8

complexity will raise exponentially as the problem size increases. Consequently, different control strategies such as numerical continuation (Ohtsuka 2004), event-triggered control (Li and Shi 2014), off-line precomputation (Büskens and Maurer 2000), and distributed implementation (Camponogara et al. 2002; Christofides et al. 2013) have been proposed attempting to shorten the computing time. Although the LMPC framework proposed in Chap. 4 admits the trade-off between computational complexity and control performance, which partially alleviates the heavy burden by circumventing this technical issue, it is desired to develop new tools to explicitly reduce the complexity of the control algorithms.

To move one step further toward real AUV applications, in this chapter, we particularly focus on the computational issue and propose two distinct implementation strategies to reduce the computational complexity of the NMPC-based AUV trajectory-tracking control algorithm while maintaining the comparable control performance. The first strategy is based on Ohtsuka's C/GMRES algorithm (Ohtsuka 2004). The C/GMRES algorithm is appealing since it entertains both merits of the numerical continuation method (Allgower and Georg 2012) and the Krylov subspace method (Kelley 1995). Applying the numerical continuation method, the NLP needs not to be solved. Instead, an approximate solution can be traced from the previous value using the solution of a linear equation. The linear equation is solved by the Krylov subspace method which is well-known for its high efficiency: The solution typically converges in a few iterations. However, the original C/GMRES algorithm was proposed for the equality-constrained NMPC problems. To incorporate the AUV thrust limits (which are essentially inequality constraints) in the NMPC formulation, we borrow barrier functions from the interior-point method and modify the C/GMRES algorithm. The sufficiency of PMP points to solve the KKT conditions is exploited. As a result, under differentiability and constraint qualifications, the validity of the modification can be claimed. Since the linear independence constraint qualification (LICQ) is satisfied in the AUV tracking control problem, the proposed modified C/GMRES algorithm provides a viable solution with high efficiency.

The second strategy exploits the dynamic properties of the AUV motion and solves the OCPs in a distributed fashion. To facilitate the distributed implementation of the NMPC tracking control, a reference augmentation is performed. With the reference augmentation, the interaction between surge, sway, and yaw dynamics is weakened. By exploiting this property, two novel distributed implementation algorithms are developed in an attempt to relieve the computation burden. By solving three subproblems with smaller sizes, the computational complexity drops significantly. The proposed parallel implementation minimizes the computational complexity while the sequential implementation sacrifices some efficiency for the guaranteed stability. Recursive feasibility and closed-loop stability are rigorously proved. The warm start strategy is used to accelerate the convergence.

The main contributions of this chapter are summarized as follows:

- The modified C/GMRES algorithm is proposed to implement the NMPC for AUV trajectory-tracking control tasks with significantly improved efficiency.

- The legitimacy of incorporating barrier functions into the numerical continuation approximations is proved based on the internal relationship between the PMP and the KKT conditions.
- Three novel distributed implementation algorithms are developed for the NMPC tracking control. The computational complexity can be dramatically reduced by adopting the developed distributed implementation methods while the tracking performance can be well maintained.
- The closed-loop properties are explicitly analyzed for the distributed implementation. Sufficient conditions to guarantee the recursive feasibility and the closed-loop stability are provided.
- Extensive simulation studies reveal the prominent robustness of the proposed fast NMPC algorithms.

8.1.2 Chapter Organization

The remaining part of this chapter is organized as follows: In Sect. 8.2, the modified C/GMRES algorithm is presented with technical details. Section 8.3 introduces a novel distributed implementation strategy and elaborates the control algorithm design and stability analysis. In Sect. 8.4, several conclusive remarks are provided.

In this chapter, the following notation is used: The column operation $[\rho_1^{\mathrm{T}}, \ldots \rho_n^{\mathrm{T}}]^{\mathrm{T}}$ is denoted as $\mathrm{col}(\rho_1, \ldots, \rho_n)$; the diagonal operation is abbreviated by $\mathrm{diag}(\cdot)$; the square of a weighted Euclidean norm $\rho^{\mathrm{T}} A \rho$ is denoted by $\|\rho\|_A^2$. The $\max\{\cdot\}$ function returns the largest value from the numbers provided in the brace. The absolute value on a vector $|\rho|$ applies the absolute value operation to each element. For a scalar-valued function f, ∇f represents the gradient while F_x denotes the Jacobian of the vector-valued function F with respect to x.

8.2 Modified C/GMRES Algorithm

8.2.1 Problem Formulation

The AUV model studied for the trajectory-tracking control is established using the kinematic equations and the dynamic equations:

$$\dot{\mathbf{x}} = \begin{bmatrix} \mathbf{R}(\psi)\mathbf{v} \\ \mathbf{M}^{-1}(\mathbf{u} - \mathbf{C}(\mathbf{v})\mathbf{v} - \mathbf{D}(\mathbf{v})\mathbf{v} - \mathbf{g}(\eta)) \end{bmatrix} = \mathbf{f}(\mathbf{x}, \mathbf{u}), \tag{8.1}$$

where the state vector $\mathbf{x} = [x, y, \psi, u, v, r]^{\mathrm{T}}$ is consisted of the pose and velocity of the vehicle, and the control vector $\mathbf{u} = [F_u, F_v, F_r]^{\mathrm{T}}$ is the generalized thrust forces and moments. The detailed expression can be found in (2.32) and (2.34).

Consider a reference trajectory $p(t) = [x_d(t), y_d(t)]^{\mathrm{T}}$ which defines the desired positions for the vehicle. We assume that the reference trajectory is appropriate by considering physical limits of a vehicle. We view $p(t)$ as the output trajectory of a reference system which owns the same kinematic properties

$$\begin{aligned} \dot{x}_d &= u_d \cos\psi_d - v_d \sin\psi_d, \\ \dot{y}_d &= u_d \sin\psi_d + v_d \cos\psi_d, \\ \dot{\psi}_d &= r_d. \end{aligned} \tag{8.2}$$

Then the kinematic equations in (8.2) can be used as a guideline to augment $p(t)$. A valid reference system for Falcon is $\mathbf{x}_d = [x_d, y_d, \psi_d, u_d, v_d, r_d]^{\mathrm{T}}$ with

$$\begin{cases} \psi_d = \operatorname{atan2}(\dot{y}_d, \dot{x}_d), \\ u_d = \sqrt{\dot{x}_d^2 + \dot{y}_d^2}, \\ v_d = 0, \\ r_d = (\dot{x}_d \ddot{y}_d - \dot{y}_d \ddot{x}_d)/(\dot{x}_d^2 + \dot{y}_d^2), \end{cases} \tag{8.3}$$

where atan2 is the four-quadrant inverse tangent function. In this way, potential numerical difficulties associated with positive semi-definite weighting matrices induced singularity or rank deficiency (when the tracking control is formulated as an output MPC problem) can be avoided. This is important because in the following we will take advantage of the numerical continuation to approximate the optimal control signal, in which the regularity of the solution, or equivalently, the local full rank of the Hessian of the Hamiltonian is required by implicit function theorem (Allgower and Georg 2012). Now we define the optimal control problem (P_0) that needs to be solved online:

$$\begin{aligned} (\mathrm{P}_0): \min_{\hat{\mathbf{u}}(s,t_0)} \ & J = \int_0^T (\|\hat{\mathbf{x}}(s,t_0) - \mathbf{x}_d(t_0+s)\|_Q^2 + \|\hat{\mathbf{u}}(s,t_0)\|_R^2) ds + \|\hat{\mathbf{x}}(T,t_0) - \mathbf{x}_d(t_0+T)\|_{Q_f}^2 \\ \text{s.t.} \ & \dot{\hat{\mathbf{x}}}(s,t_0) = \mathbf{f}(\hat{\mathbf{x}}(s,t_0), \hat{\mathbf{u}}(s,t_0)), \\ & \hat{\mathbf{x}}(0,t_0) = \mathbf{x}(t_0), \\ & |\hat{\mathbf{u}}(s,t_0)| \le \mathbf{u}_{\max}, \end{aligned} \tag{8.4}$$

where $\hat{\mathbf{x}}(s, t_0)$ denotes the predicted state evolution starting from $\mathbf{x}(t_0)$ at the origin of the fictitious time axis $s = 0$. The optimal control signal $\hat{\mathbf{u}}^*(s, t_0)$ is determined by solving the above optimization. T is the prediction horizon, and Q_f, Q, R are positive definite weighting matrices.

The NMPC algorithm for the AUV tracking can be briefly described as follows:

- At sampling instant t_0, the optimization problem (P_0) is performed with the current system state $\mathbf{x}(t_0)$, and let $\hat{\mathbf{u}}^*(s, t_0)$ denote the solution.
- The AUV uses $\hat{\mathbf{u}}^*(s, t_0)$ for only one sampling period, that is, $\mathbf{u}(t) = \hat{\mathbf{u}}^*(s, t_0)$ for $s \in [0, \Delta t]$.
- At next sampling instant $t_0 + \Delta t$, new measurement of system state $\mathbf{x}(t_0 + \Delta t)$ is fed back, then (P_0) is solved again with t_0 substituted by $t_0 + \Delta t$.

The above procedure will be repeated until accomplishing the tracking task.

We notice that in (P_0) the objective function J includes an integral operation and the constraints include the control system dynamics which are essentially derivative operations. However, in real-world implementation, both the integral and the derivative have to be performed numerically. Therefore, we solve the discretized version of (P_0) at each sampling instant.

For the convenience of the description of C/GMRES algorithm, let us slightly abuse the notation. In the following, u represents the control input of a general nonlinear dynamical system $\dot{x} = f(x, u, p)$ rather than AUV surge velocity; x represents the state rather than the AUV position; and p is the time-varying parameter. We generalize (P_0) and the discretized version (P) using Euler forward approximation as follows:

$$
\begin{aligned}
(\mathrm{P}) : \min_{u_i} . \; & J = \sum_{i=0}^{N-1} \ell(x_i, u_i, p_i)\Delta t + g(x_N, p_N) \\
\text{s.t.} \quad & x_{i+1} = x_i + f(x_i, u_i, p_i)\Delta t, \\
& x_0 = x(t_0), \\
& c(x_i, u_i, p_i) = 0, \\
& h(x_i, u_i, p_i) \leq 0.
\end{aligned}
\tag{8.5}
$$

Here, we divide the prediction horizon T into N steps with step size Δt (also known as the sampling period). The continuous-time system dynamics $\dot{x} = f(x, u, p)$ is also discretized, and the initial condition is given by the current system state $x(t_0)$. $c(x_i, u_i, p_i), h(x_i, u_i, p_i)$ represent vector-valued equality and inequality constraints, respectively.

8.2.2 Solving the NMPC Problem

Generally, solving the optimization problem (P) is a NLP problem. Under differentiability and constraint qualifications, the KKT theorem can be applied. The KKT theorem states the first-order necessary conditions for a solution $\{u_i^*\}_{i=0}^{N-1}$ of (P) to be optimal (Andreas and Lu 2007). Since the system dynamics and initial condition of (P) are essentially equality constraints, we combine them with rest of the equalities and denote them by $\bar{c}(x_i, u_i, p_i) = 0$. We simplify the notation of (P) as follows:

$$
\min_{\bar{U}} . \; J(\bar{U}) \;\; \text{s.t.} \; \bar{c}(\bar{U}) = 0, \;\; h(\bar{U}) \leq 0,
$$

where $\bar{U} = \mathrm{col}(u_0, u_1, \ldots, u_{N-1})$. The KKT theorem states: For a locally optimal solution $\bar{U}^*$, there exist vector-valued multipliers λ^* and μ^* such that the following conditions hold:

$$
\begin{aligned}
&\nabla J(\bar{U}^*) + \textstyle\sum_i \lambda_i^* \nabla \bar{c}_i(\bar{U}^*) + \sum_j \mu_j^* \nabla h_j(\bar{U}^*) = 0, \\
&\bar{c}_i(\bar{U}^*) = 0, \\
&h_j(\bar{U}^*) \le 0, \\
&\mu_j^* \ge 0, \\
&\mu_j^* h_j(\bar{U}^*) = 0.
\end{aligned} \tag{8.6}
$$

To solve (P) is actually to solve the KKT system (8.6) using Newton's methods such as the SQP method and the interior-point (IP) method (Andreas and Lu 2007). However, conventional Newton-type numerical algorithms basically rely on successive linearizations, essentially making it computationally expensive in solving the problem. To alleviate the computational burden, many efficient approximate algorithms for MPC have been developed (Diehl et al. 2009). For the AUV tracking control, Ohtsuka's C/GMRES algorithm (Ohtsuka 2004) is tailored and modified to incorporate the physical limits of the thrusters.

8.2.3 Modified C/GMRES Algorithm

Instead of solving the generic NMPC problem (P), the original C/GMRES algorithm considers the less general problem (P_1) in which no inequality constraints exist

$$
\begin{aligned}
(\mathrm{P}_1): \min_{u_i}.\ & J = \sum_{i=0}^{N-1} \ell(x_i, u_i, p_i)\Delta t + g(x_N, p_N) \\
\text{s.t.}\quad & x_{i+1} = x_i + f(x_i, u_i, p_i)\Delta t, \\
& x_0 = x(t_0), \\
& c(x_i, u_i, p_i) = 0.
\end{aligned}
$$

The C/GMRES is based on solving the discretized necessary condition for the optimal control, i.e., the PMP (Ohtsuka 2004). Define the Hamiltonian by

$$
H(x, \bar{\lambda}, u, \nu, p) = \ell(x, u, p) + \bar{\lambda}^{\mathrm{T}} f(x, u, p) + \nu^{\mathrm{T}} c(x, u, p).
$$

Here, we call $\bar{\lambda}$ the costate and ν the language multiplier. The discretized PMP claims that for a local optimal control $\{u_i^*\}_{i=0}^{N-1}$, there exist $\{\bar{\lambda}_i^*\}_{i=0}^{N}$ and $\{\nu_i^*\}_{i=0}^{N-1}$ satisfying the following conditions:

$$x_{i+1}^* = x_i^* + f(x_i^*, u_i^*, p_i)\Delta t, \tag{8.7a}$$

$$\bar{\lambda}_i^* = \bar{\lambda}_{i+1}^* + H_x^{\mathrm{T}}(x_i^*, \bar{\lambda}_{i+1}^*, u_i^*, \nu_i^*, p_i)\Delta t, \tag{8.7b}$$

$$\bar{\lambda}_N^* = g_x^{\mathrm{T}}(x_N^*, p_N), \tag{8.7c}$$

$$x_0^* = x(t_0), \tag{8.7d}$$

$$H_u(x_i^*, \bar{\lambda}_{i+1}^*, u_i^*, \nu_i^*, p_i) = 0, \tag{8.7e}$$

$$c(x_i^*, u_i^*, p_i) = 0. \tag{8.7f}$$

Observe that (8.7a)–(8.7b) are recurrence relations on x^* and $\bar{\lambda}^*$, and (8.7c)–(8.7d) can be viewed as boundary conditions. Define $U = \mathrm{col}(u_0^*, \nu_0^*, \ldots, u_{N-1}^*, \nu_{N-1}^*)$, then the state trajectory $\{x_i^*\}_{i=0}^N$ and the costate trajectory $\{\bar{\lambda}_i^*\}_{i=0}^N$ are explicit functions of U. We substitute (8.7a)–(8.7d) into (8.7e)–(8.7f) and construct the equation system

$$F(U, x) = \begin{bmatrix} H_u^{\mathrm{T}}(x_0^*, \bar{\lambda}_1^*, u_0^*, \nu_0^*, p_0) \\ c(x_0^*, u_0^*, p_0) \\ \vdots \\ H_u^{\mathrm{T}}(x_{N-1}^*, \bar{\lambda}_N^*, u_{N-1}^*, \nu_{N-1}^*, p_{N-1}) \\ c(x_{N-1}^*, u_{N-1}^*, p_{N-1}) \end{bmatrix} = 0. \tag{8.8}$$

Normally, we need to solve (8.8) at each sampling instant. However, solving (8.8) using iterative methods is computationally expensive in the sense of evaluating Jacobians, Hessian, and inverses. Instead, C/GMRES takes the advantage of numerical continuation method (Allgower and Georg 2012) that views the equation system $F(U, x)$ as a dynamical system governed by

$$\dot{F}(U, x, t) = A_s F(U, x, t). \tag{8.9}$$

Here, we explicitly point out the time dependence of the dynamical system (8.9) by denoting the state as $F(U, x, t)$ rather than solely $F(U, x)$. A_s is an introduced stable matrix to stabilize $F(U, x, t)$ at the origin. Hence, if F_U is nonsingular, (8.9) is equivalent to

$$\dot{U} = F_U^{-1}(A_s F - F_x \dot{x} - F_t). \tag{8.10}$$

The derivative of U can be computed by (8.10) using the derivative of x. Then, if an initial solution $U(0)$ satisfying $F(U(0), x(0), 0) = 0$ can be found, the solution curve $U(t)$ of (8.8) can be traced by integrating (8.10) from $U(0)$.

To further relieve the computational load, the action of Jacobians on vectors is approximated by forward difference:

$$\begin{aligned} &F_U(U, x, t)W + F_x(U, x, t)w + F_t(U, x, t)\omega \\ &\approx \delta^{-1}(F(U + \delta W, x + \delta w, t + \delta\omega) - F(U, x, t)) \\ &= D_\delta F(U, x, t : W, w, \omega), \end{aligned}$$

where δ is a small positive number. To further avoid the expensive operation of matrix inverse F_U^{-1}, generalized minimal residual method (GMRES) is applied to solve the linear equation $F_U \dot{U} = AsF - F_x \dot{x} - F_t$. The combination of forward difference approximation and GMRES is called FDGMRES introduced in Kelley (1995). To simplify the description, we can simply view FDGMRES as a function

$$\dot{U} = \mathrm{FDGMRES}(U, x, \dot{x}, t, \hat{\dot{U}}, \delta, k_{\max}),$$

where $\hat{U}$ is an initial guess and $k_{\max}$ is the allowed maximum iteration number. By defining a transformation $T_0(U) = u_0^*$, the C/GMRES algorithm can be depicted in Algorithm 8.1.

Algorithm 8.1 C/GMRES Algorithm

1: Initialize $t = 0, k = 0$, initial state $x_0 = x(0)$ and find U_0 numerically such that $F(U_0, x_0, 0) = 0$.
2: For $s \in [t, t + \Delta t)$, the real control input is computed by $u(s) = T_0(U_k)$.
3: At next sampling instant $t + \Delta t$, the system state $x_{k+1} = x(t + \Delta t)$ is fed back. Compute the state difference $\Delta x_k = x_{k+1} - x_k$.
4: $\dot{U}_k$=FDGMRES($U_k, x_k, \Delta x_k/\Delta t, t, \hat{\dot{U}}_k, \delta, k_{\max}$)
5: Set $U_{k+1} = U_k + \dot{U}_k \Delta t$
6: Update $t = t + \Delta t, k = k + 1$
7: Go to Step 2.

Since several approximations are made in the C/GMRES algorithm, the approximation error should be bounded. An error analysis can be found in Ohtsuka (2004).

The AUV tracking problem (P_0), however, is constrained by inequality constraints in terms of actuator limits. We modify the C/GMRES algorithm appropriately so that it can incorporate the inequality constraints. By introducing the barrier functions (Boyd et al. 2004), we approximately solve (P) by actually solving another optimization ($\bar{\mathrm{P}}$) defined as follows:

$$(\bar{\mathrm{P}}) : \min_{\bar{U}} \; J(\bar{U}) - \gamma_l \textstyle\sum_j \log(-h_j(\bar{U})) \;\; \text{s.t.} \; \bar{c}(\bar{U}) = 0, \tag{8.11}$$

where γ_l is a positive number. Since the C/GMRES is based on solving PMP system rather than KKT system, in the following, we first explore the sufficiency of PMP points to satisfy the KKT conditions, then exploit the convergence of the solution of ($\bar{\mathrm{P}}$) to that of (P).

Lemma 8.1 *A point that satisfies the PMP of (P_1) sufficiently solves the corresponding KKT conditions.*

Proof The detailed KKT conditions of (P_1) are derived in the following:

$$\nabla \bar{J} = 0, \tag{8.12a}$$

$$\bar{c}(x_i, u_i, p_i) = 0, \tag{8.12b}$$

with

$$\begin{aligned} \bar{J} = &\textstyle\sum_{i=0}^{N-1} L(x_i, u_i, p_i) + g(x_N, p_N) + \sum_{i=0}^{N-1} \bar{\lambda}_{j+1}^{\mathrm{T}}(f_d(x_j, u_j, p_j) \\ &- x_{j+1}) + \tilde{\lambda}_0^{\mathrm{T}}(x(t_0) - x_0) + \textstyle\sum_{k=0}^{N-1} \nu_k^{\mathrm{T}} c(x_k, u_k, p_k)\Delta t, \end{aligned}$$

where $L(x_i, u_i, p_i) = \ell(x_i, u_i, p_i)\Delta t$ and $f_d(x_i, u_i, p_i) = x_i + f(x_i, u_i, p_i)\Delta t$. Define the Hamiltonian sequence by

$$H^i = L(x_i, u_i, p_i) + \bar{\lambda}_{i+1}^{\mathrm{T}} f_d(x_i, u_i, p_i) + \nu_i^{\mathrm{T}} c(x_i, u_i, p_i)\Delta t.$$

Then we can simplify the notation of $\bar{J}$

$$\bar{J} = g(x_N, p_N) - \bar{\lambda}_N^{\mathrm{T}} x_N + \sum_{i=1}^{N-1} (H^i - \bar{\lambda}_i^{\mathrm{T}} x_i) + H^0 + \tilde{\lambda}_0^{\mathrm{T}}(x(t_0) - x_0).$$

Since given x_0, the system state x_i is merely dependent on $\bar{U}$, the gradient

$$\nabla \bar{J} = \frac{\partial \bar{J}}{\partial X}\frac{dX}{d\bar{U}} + \frac{\partial \bar{J}}{\partial \bar{U}},$$

where $X = \mathrm{col}(x_0, \ldots, x_N)$, then $\nabla \bar{J} = 0$ is equivalent to

$$d\bar{J} = \frac{\partial \bar{J}}{\partial X} dX + \frac{\partial \bar{J}}{\partial \bar{U}} d\bar{U} = 0.$$

By explicitly computing $\mathrm{d}\bar{J}$, we have

$$\begin{aligned} \mathrm{d}\bar{J} = &\left(\tfrac{\partial g}{\partial x_N} - \bar{\lambda}_N^{\mathrm{T}}\right) \mathrm{d}x_N + \sum_{i=1}^{N-1} \{\left(\tfrac{\partial H^i}{\partial x_i} - \bar{\lambda}_i^{\mathrm{T}}\right) \mathrm{d}x_i \\ &+ \tfrac{\partial H^i}{\partial u_i} \mathrm{d}u_i\} + \tfrac{\partial H^0}{\partial x_0} \mathrm{d}x_0 + \tfrac{\partial H^0}{\partial u_0} \mathrm{d}u_0 - \tilde{\lambda}_0^{\mathrm{T}} \mathrm{d}x_0 = 0. \end{aligned} \tag{8.13}$$

If we select the costate sequence satisfying

$$\bar{\lambda}_i^{\mathrm{T}} - \frac{\partial H^i}{\partial x_i} = 0, \tag{8.14a}$$

$$\frac{\partial g}{\partial x_N} - \bar{\lambda}_N^{\mathrm{T}} = 0, \tag{8.14b}$$

and $x_0 = x(t_0)$ is given, then $d\bar{J} = 0$ is equivalent to

$$\mathrm{d}\bar{J} = \sum_{i=0}^{N-1} \frac{\partial H^i}{\partial u_i} \mathrm{d}u_i = 0. \tag{8.15}$$

We notice that (8.14a)–(8.14b) are identical to (8.7b)–(8.7c) and (8.7e) sufficiently solves (8.15). Equation (8.12b) states the rest of PMP conditions. Now, we can find that a feasible point that satisfies the PMP conditions sufficiently solves the KKT conditions. □

Lemma 8.2 *Suppose differentiability and constraint qualifications hold. Let a sequence* $\{\gamma_l\}$ *satisfy* $0 < \gamma_{l+1} < \gamma_l$*, and* $\gamma_l \to 0$ *as* $l \to \infty$*. Suppose the solution of* (P) *and* ($\bar{P}$) *exist and denoted by* $\bar{U}^*$ *and* $\bar{U}^l$*, then* $\bar{U}^l$ *converge to* $\bar{U}^*$ *as* $\gamma_l \to 0$.

Proof See Boyd et al. (2004) [Chap. 11, Sect. 3.3]. □

Definition 8.1 (*Linear Independence Constraint Qualification*) Let $\bar{U}$ be feasible for (P) and $I(\bar{U}) = \{j | h_j(\bar{U}) = 0\}$, then we say that the linear independence constraint qualification (LICQ) holds at $\bar{U}$, if for $j \in I(\bar{U})$

$$\nabla h_j(\bar{U}), \nabla \bar{c}_i(\bar{U}),$$

are linearly independent. Then strong duality gap holds for (P) and KKT theorem is applicable.

Theorem 8.1 *For the AUV tracking problem (8.4), the modified C/GMRES with log barriers approximately solves* (P_0) *with a small positive* γ_l.

Proof Since the system dynamics $\dot{\mathbf{x}}(t) = \mathbf{f}(\mathbf{x}(t), \mathbf{u}(t))$ imposes constraints on $\mathbf{x}$ not on $\mathbf{u}$, the only constraints on $\mathbf{u}$ are $|\mathbf{u}(t)| \leq \mathbf{u}_{\max}$. For the discretized version, the constraints are $|\mathbf{u}_i| \leq \mathbf{u}_{\max}$, or equivalently, $\mathbf{u}_i \leq \mathbf{u}_{\max}$ and $\mathbf{u}_i \geq -\mathbf{u}_{\max}$, which obviously satisfy LICQ at all feasible points. Therefore, Theorem 8.1 is followed by Lemmas 8.1 and 8.2. □

The modified C/GMRES algorithm can be depicted in Algorithm 8.2.

Algorithm 8.2 mC/GMRES Algorithm

1: Initialize $t = 0$, $k = 0$ and $l = 0$. Given initial state $x_0 = x(0)$, γ_0, α and $l_{\max}$, find U_0^l numerically such that $F(U_0^l, x_0, \gamma_l, 0) = 0$.
2: Set $l = l + 1$. If $l < l_{\max}$, go to Step 3. Otherwise, go to Step 4.
3: Set $\gamma_l = \alpha\gamma_{l-1}$. Then find U_0^l numerically such that $F(U_0^l, x_0, \gamma_l, 0) = 0$ with initial guess $\hat{U}_0^l = U_0^{l-1}$. Go to Step 2.
4: For $s \in [t, t + \Delta t)$, the real control input is computed by $u(s) = T_0(U_k)$.
5: At next sampling instant $t + \Delta t$, the system state $x_{k+1} = x(t + \Delta t)$ is fed back. Compute the state difference $\Delta x_k = x_{k+1} - x_k$.
6: $\dot{U}_k$=FDGMRES($U_k, x_k, \Delta x_k/\Delta t, t, \hat{\dot{U}}_k, \delta, k_{\max}$)
7: Set $U_{k+1} = U_k + \dot{U}_k \Delta t$
8: Update $t = t + \Delta t$, $k = k + 1$
9: Go to Step 4.

Note that the conventional C/GMRES algorithm only solves the PMP system (8.7) once at the initialization, and the modified version changes the initialization by solving $F(U_0, x_0, \gamma_l, 0) = 0$ several times with a shrinking γ_l and initial guesses $\hat{U}_0^{l+1} = U_0^l$ in order to obtain good approximations while circumventing numerical difficulties (Boyd et al. 2004). Therefore, the modification preserves the efficiency of the fast algorithm.

8.2.4 Simulation Study

In this section, the AUV tracking control with three different reference trajectories is simulated. We demonstrate the effectiveness and highlight the efficiency of the modified C/GMRES by comparisons to other numerical algorithms. All the simulations are performed on a personal computer (CPU: Intel(R) Core(TM) i7-3520M: 2.90GHz 2.90GHz; RAM: 4.00GB).

The first reference trajectory (Case I) to be tested is a circle defined as follows:

$$p(t) = \begin{cases} x_d = 0.8\cos(0.5t), \\ y_d = 0.8\sin(0.5t). \end{cases} \tag{8.16}$$

The circle trajectory is a typical test trajectory for tracking control problems. It is relative easier to track because the reference angular and tangential velocities keep constant values.

The second test reference trajectory (Case II) is a sinusoidal curve defined as follows:

$$p(t) = \begin{cases} x_d = 0.5t, \\ y_d = \sin(0.5t). \end{cases} \tag{8.17}$$

The sinusoidal trajectory is another typical test trajectory for tracking control whose reference tangential and angular velocities are no longer constant.

The third test reference trajectory (Case III) is an eight-shaped trajectory defined as follows:

$$p(t) = \begin{cases} x_d = \sin(0.5t), \\ y_d = \sin(0.25t). \end{cases} \tag{8.18}$$

The eight-shaped trajectory has sharp changes in its reference velocities, which makes the tracking task more challenging.

We notice that in the augmented reference system (8.3), the first and second derivatives of $p(t)$ are needed. It could be tedious to analytically calculate these functions. Instead, in the implementation, we approximate the discrete values by forward differences:

$$\begin{aligned} \dot{p}(n\Delta t) &\approx \frac{p((n+1)\Delta t) - p(n\Delta t)}{\Delta t}, \\ \ddot{p}(n\Delta t) &\approx \frac{p((n+2)\Delta t) - 2p((n+1)\Delta t) + p(n\Delta t)}{(\Delta t)^2}. \end{aligned}$$

Note that the above process will introduce some approximation errors, so there exists a trade-off between computational complexity and solution accuracy.

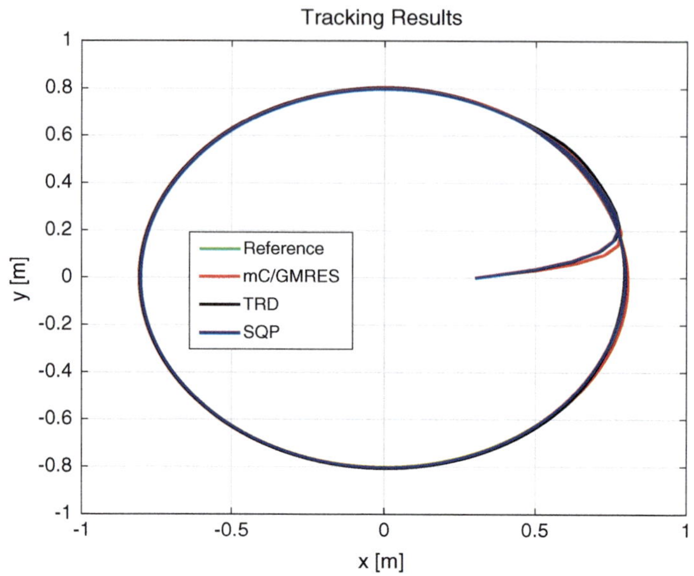

Fig. 8.1 Simulated AUV trajectories—Case I

The AUV system parameters are from the identified dynamic model of Falcon (Alison 2014). The model parameter details are summarized in Table 2.1. Thrust limits $F_{u,\max} = 500$ (N), $F_{v,\max} = 500$ (N), and $F_{r,\max} = 500$ (Nm) and initial conditions $\mathbf{x}(0) = [0.3, 0, 0, 0, 0, 0]^{\mathrm{T}}$. For the NMPC parameters, sampling period $\Delta t = 0.1$ s, weighting matrices $Q = \mathrm{diag}(10^4, 10^4, 10, 10, 1, 10)$, $R = \mathrm{diag}(10^{-4}, 10^{-4}, 10^{-2})$, and $Q_f = \mathrm{diag}(10, 10, 1, 1, 1, 1)$. For the modified C/GMRES algorithm, $A_s = -\zeta I$ with $\zeta = 1/\Delta t$, $k_{\max} = 4$, $\delta = 10^{-5}$. In the initialization $\gamma_0 = 10^2$ and iteratively solve $F(U_0, x_0, \gamma_l, 0) = 0$ with $\gamma_{l+1} = \alpha\gamma_l = \gamma_l/10$ and $\hat{U}_0^{l+1} = U_0^l$ until $l = l_{\max} = 6$.

In all of the cases, at each sampling instant, the optimal control signal of the NMPC tracking is computed by (i) the proposed modified C/GMRES algorithm (red curves), (ii) the trust–region–dogleg (TRD) algorithm (Yuan 2000) to solve $F(U, x, t) = 0$ (black curves), and (iii) the embedded SQP (Andreas and Lu 2007) of MATLAB function `fmincon` to directly solve the KKT system of (P_0) (blue curves), respectively. The green curves are the reference trajectories $p(t)$.

The NMPC tracking control results of a circle are shown in Fig. 8.1. The red, black, and blue curves are simulated AUV trajectories. The control forces and moments are recorded in Fig. 8.2, and the position errors are plotted in Fig. 8.3 with respect to time. Generally speaking, the AUV tracking control is comparably well with different implementations. Position errors are bounded. Also, as expected, the control signals are within their intervals of permitted values, which validates the usefulness of barrier functions.

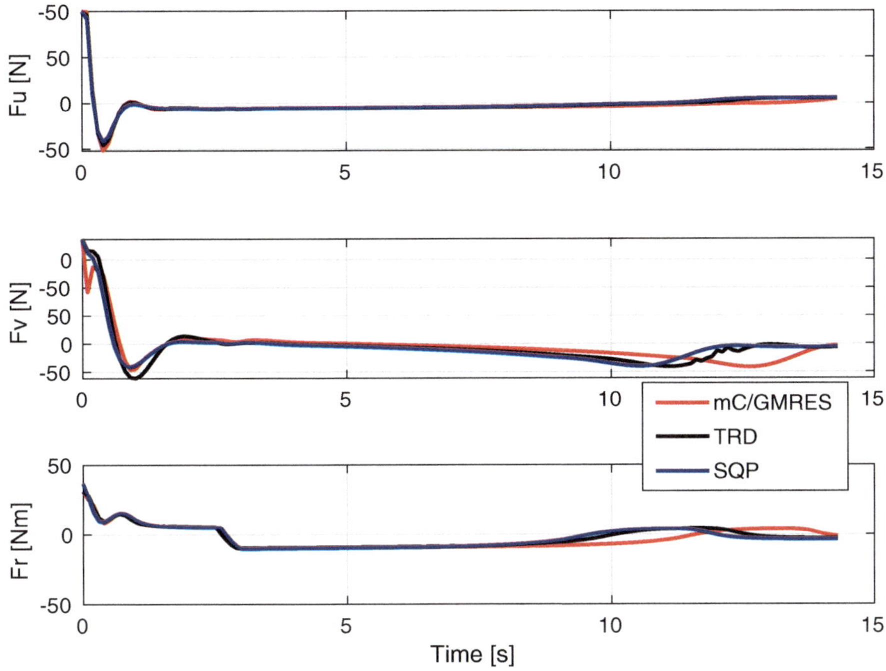

Fig. 8.2 Control forces and moments—Case I

The simulated AUV trajectories to track a sinusoidal reference are illustrated in Fig. 8.4. The control forces and moments are plotted in Fig. 8.5, and the position errors are recorded in Fig. 8.6. The same as Case I, we can observe that the AUV tracking performance is comparably well with the three algorithms, and the control signals never go beyond their limits.

The simulation results of tracking the eight-shape reference trajectory are demonstrated in Fig. 8.7. The control signals are shown in Fig. 8.8, and the position errors are plotted in Fig. 8.9. Similar observations can be made that the NMPC tracking control seems comparably well using different numerical algorithms. Position errors are bounded, and the control signals never transcend the boundary.

Although the tracking control performance is comparably acceptable with different implementation, if we check the average computation time, as listed in Tables 8.1 8.2 and 8.3, we can easily distinguish the efficiencies of these algorithms. Remember that, the sampling period $\Delta t = 0.1$ s, which means at each sampling instant the optimal control signals have to be obtained within 0.1 s. Obviously, the embedded SQP using conventional Newton failed to solve (8.6) within the time constraint, and TRD only marginally satisfied the time constraint with small prediction horizons. In contrast, modified C/GMRES efficiently solved the NMPC problem within 10% of Δt, which is very promising for the real-time implementation of NMPC-based AUV tracking control.

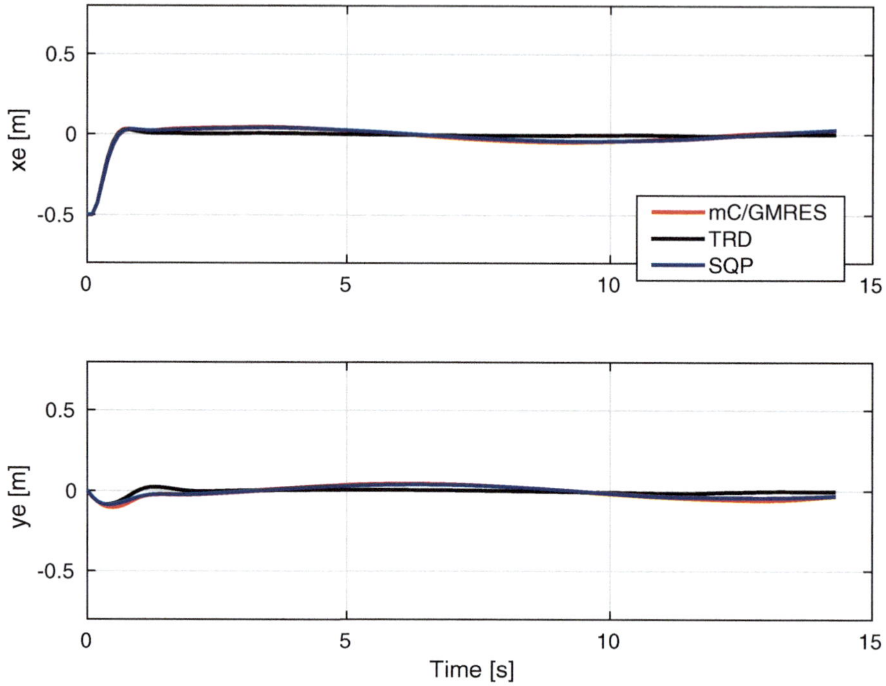

Fig. 8.3 Position error—Case I

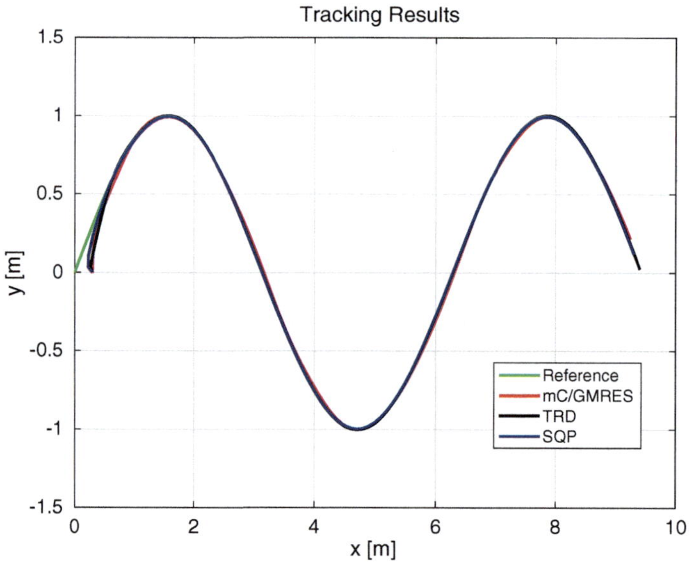

Fig. 8.4 Simulated AUV trajectories—Case II

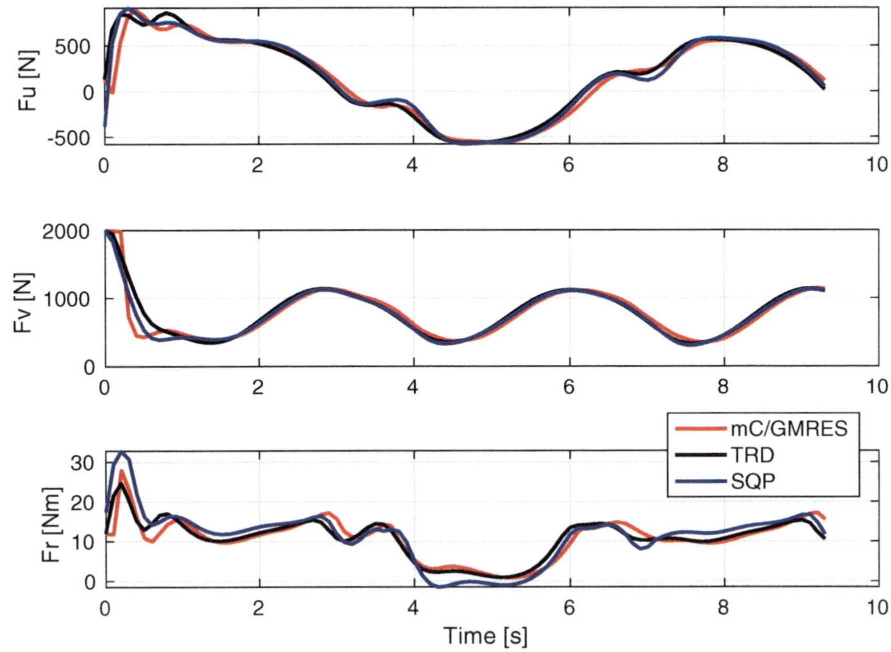

Fig. 8.5 Control forces and moments—Case II

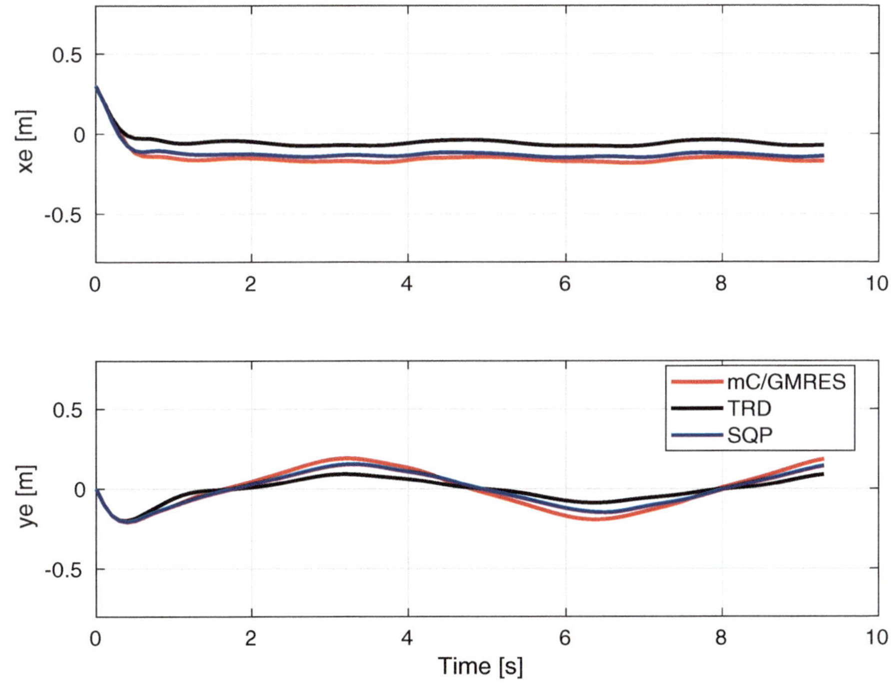

Fig. 8.6 Position error—Case II

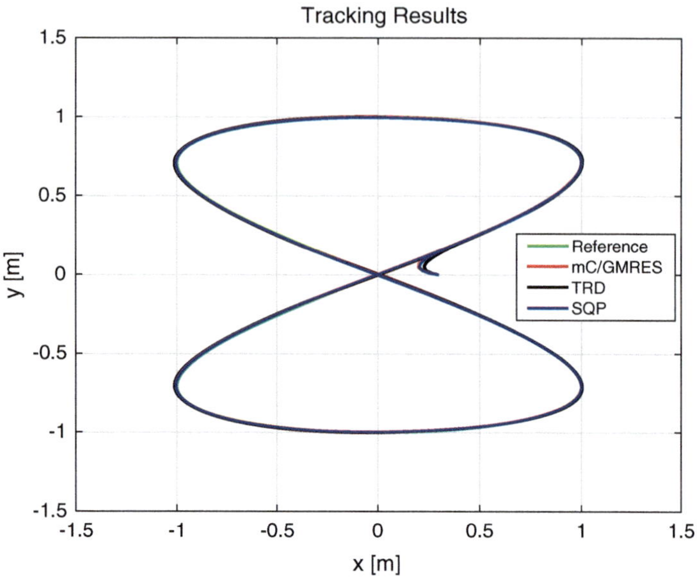

Fig. 8.7 Simulated AUV trajectories—Case III

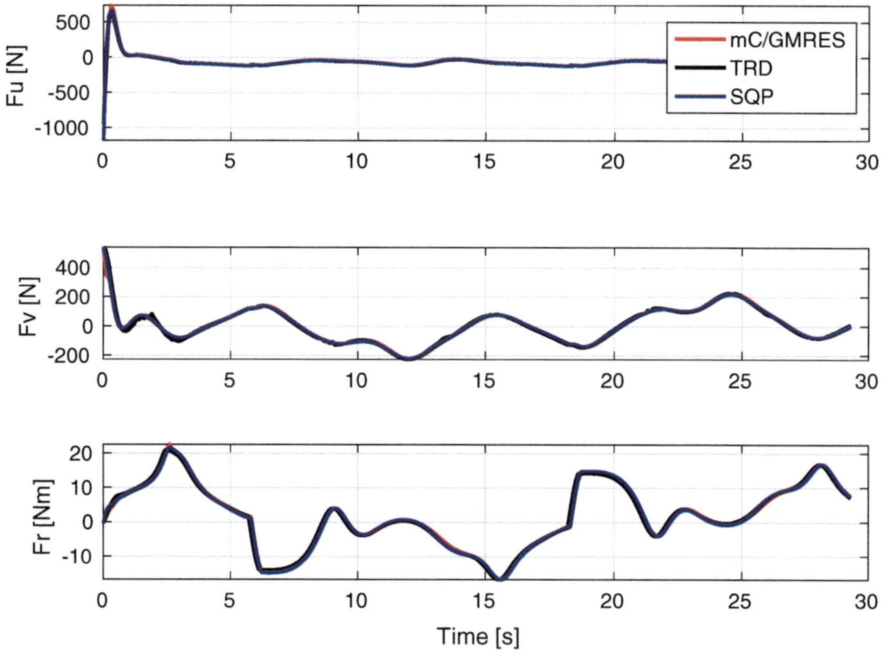

Fig. 8.8 Control forces and moments—Case III

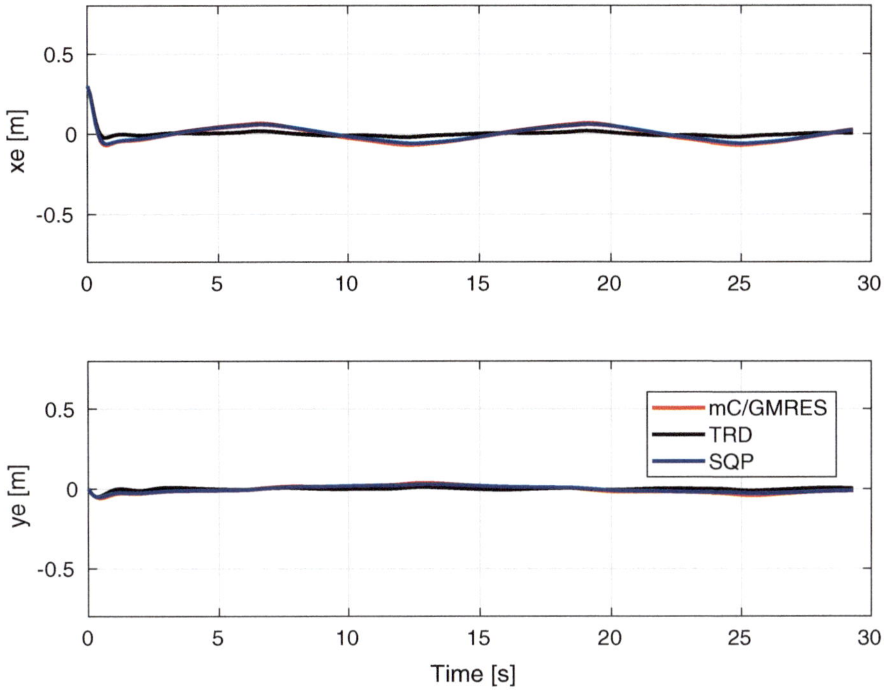

Fig. 8.9 Position error—Case III

Table 8.1 Average computation time (s) per update—Case I

N	mC/GMRES	Trust–Region–Dogleg	SQP
5	0.0038	0.0274	0.1741
10	0.0058	0.0564	0.5881
20	0.0098	0.2168	2.0969

Table 8.2 Average computation time (s) per update—Case II

N	mC/GMRES	Trust–Region–Dogleg	SQP
5	0.0033	0.0276	0.2395
10	0.0055	0.0927	0.5523
20	0.0097	0.2507	1.8901

Table 8.3 Average computation time (s) per update—Case III

N	mC/GMRES	Trust–Region–Dogleg	SQP
5	0.0040	0.0309	0.1990
10	0.0064	0.0605	0.6167
20	0.0100	0.2218	2.2646

8.3 Distributed Implementation Strategy

In the previous section, we have witnessed the dramatically improved numerical efficiency of the proposed modified C/GMRES algorithm. Due to the numerical continuation step involved in the algorithm, the optimal solution at each sampling time instant can be traced from the previous solution without truly solving the NLP problem. On one side, this can be extremely efficient in obtaining the optimal control signal. On the flip side, however, the approximation error will accumulate over time. The large approximation error may lead to highly degraded control performance or even instability. Therefore, in this section, we are going to take another fast implementation strategy which reduces the overall computational complexity while preserving the closed-loop stability.

8.3.1 Problem Formulation

The AUV model considered in this section is established based on the kinematic equations, the dynamic equations, and the thrust distribution:

$$\dot{\mathbf{x}} = \begin{bmatrix} \mathbf{R}(\psi)\mathbf{v} \\ \mathbf{M}^{-1}(\mathbf{Bu} - \mathbf{C}(\mathbf{v})\mathbf{v} - \mathbf{D}(\mathbf{v})\mathbf{v} - \mathbf{g}(\eta)) \end{bmatrix} = \mathbf{f}(\mathbf{x}, \mathbf{u}), \tag{8.19}$$

where the state vector $\mathbf{x} = [x, y, \psi, u, v, r]^{\mathrm{T}}$ is consisted of the pose and velocity of the vehicle, and the control vector $\mathbf{u} = [u_1, u_2, u_3, u_4]^{\mathrm{T}}$ is consisted of the forces generated by the four thrusters. The detailed expression can be found in (2.32) and (2.34).

The trajectory of interest is defined in the output space: $p(t) = [x_d(t), y_d(t)]^{\mathrm{T}}$ which describes the desired position in the IRF. To avoid singularities, we assume that Assumption 4.3 can hold. Then taking advantage of Proposition 4.2, we can safely formulate the NMPC tracking control problem (P_0) as follows:

$$\min_{\mathbf{u} \in S(\delta)} J = \int_0^T (\|\tilde{\mathbf{x}}(s)\|_Q^2 + \|\mathbf{u}(s)\|_R^2) ds + \|\tilde{\mathbf{x}}(T)\|_P^2 \tag{8.20a}$$

$$\text{s.t. } \dot{\mathbf{x}}(s) = \mathbf{f}(\mathbf{x}(s), \mathbf{u}(s)), \tag{8.20b}$$

$$\mathbf{x}(0) = \mathbf{x}(t_0), \tag{8.20c}$$

$$|\mathbf{u}(s)| \leq \mathbf{u}_{\max}, \tag{8.20d}$$

where $\tilde{\mathbf{x}} = \mathbf{x} - \mathbf{x}_d$ with $\mathbf{x}_d = [x_d, y_d, \psi_d, u_d, v_d, r_d]^{\mathrm{T}}$ defined in (8.3) denotes the state error; $S(\delta)$ is the family of piecewise constant functions characterized by the sampling period δ; $T = N\delta$ is known as the prediction horizon, and $Q = \text{diag}(q_{ii})$, $R = \text{diag}(r_{ii})$, $P = \text{diag}(p_{ii})$ are positive definite weighting matrices.

To establish a feedback mechanism, the NMPC tracking control needs to be implemented in the receding horizon fashion, i.e., we solve (P_0) for the current sampling time instant, and implement the solution for one sampling period, and then repeat the procedure at next sampling time. The NMPC tracking control algorithm is summarized in Algorithm 8.3.

Algorithm 8.3 Centralized Implementation Algorithm

1: Initialization: Input the weighting matrices and prediction horizon in (8.20).
2: Fetch the state measurement $\mathbf{x}(t)$.
3: Solve the NMPC problem (8.20) with $\mathbf{x}(t_0) = \mathbf{x}(t)$, and let $\mathbf{u}^*(s)$ denote the solution.
4: Use $\mathbf{u}^*(s)$ for only one sampling period: $\mathbf{u}(t) = \mathbf{u}^*(s)$ for $s \in [0, \delta]$;
5: At next sampling time, set $t = t + \delta$; Goto step 2.

8.3.2 Solving the NMPC Problem

In the NMPC problem (P_0), the performance index J includes an integral operation and the constraints encompass the system model which are essentially derivative operations. In engineering practice, both the integral and derivative operations need to be performed numerically. Equivalently, instead of solving (P_0) in continuous-time setting, we solve the discretized problem:

$$\min_{U} \ J = \sum_{k=0}^{N-1} (\|\tilde{\mathbf{x}}(k)\|_Q^2 + \|\mathbf{u}(k)\|_R^2) + \|\tilde{\mathbf{x}}(N)\|_P^2 \tag{8.21a}$$

$$\text{s.t. } \mathbf{x}(k+1) = f(\mathbf{x}(k), \mathbf{u}(k)), \tag{8.21b}$$

$$\mathbf{x}(0) = \mathbf{x}(t_0), \tag{8.21c}$$

$$|\mathbf{u}(k)| \leq \mathbf{u}_{\max}. \tag{8.21d}$$

Here, the system model is discretized and represented by $\mathbf{x}(k+1) = f(\mathbf{x}(k), \mathbf{u}(k))$ with $\mathbf{x}(i) = \mathbf{x}(i\delta)$ and $\mathbf{u}(i) = \mathbf{u}(i\delta)$; the sequence of control inputs $U = \text{col}(\mathbf{u}(0), \mathbf{u}(1), \ldots, \mathbf{u}(N-1))$ are the decision variables.

Since f is nonconvex and is imposed in the constraints, solving (8.21) is a generic NLP problem. Solving a nonconvex NLP problem is computationally expensive and unrealistic in real-time control if the problem size is too large. Therefore, it is desirable to (approximately) inspect the computational complexity. The SQP method represents one of the most widely used and effective NLP solvers, and we take SQP for example.

To solve the NLP, SQP generates a sequence of iterates which satisfies the first order KKT conditions. Each iterate is the solution of a QP subproblem that linearly approximates the local behavior of the original NLP problem in a neighborhood of the initial feasible point. Compared to the active-set counterpart which is of exponential complexity, the interior-point method has the polynomial complexity and is suggested to solve the QP subproblem. As a benchmark, the Mehrotra's predictor–corrector algorithm (Mehrotra 1992) is investigated.

The worst-case number of floating-point operations (flops) for the QP subproblem is given by the following formula (Richter et al. 2012):

$$\#\text{flops}_{\text{IP}} = i_{\text{IP}}(2/3(Nm)^3 + 2(Nm)^2), \tag{8.22}$$

for the condense formulation. Here, N is the prediction horizon, and m is the number of control inputs, hence Nm is the number of decision variables for the optimization problem; i_{IP} is the number of interior-point iterations which is expected to be $\mathcal{O}(\sqrt{Nm}\log(1/\epsilon))$ (McGovern 2000) for the ϵ-accurate solution, or simply a specified maximum iteration number $i_{\text{IP,max}}$. Neglecting minor operations in function evaluations, gradient and Hessian updating, and line search in one major SQP iteration, a rough approximation of the computational complexity for numerically solving (P_0) using SQP can be given as

$$\#\text{flops}_{\text{SQP}} \approx i_{\text{SQP}} \times (\#\text{flops}_{\text{IP}}), \tag{8.23}$$

where i_{SQP} denotes the numbers of SQP iterations which is specified by the maximum iteration number $i_{\text{SQP,max}}$. From (8.23) we know that shrinking the sizes of m and N effectively reduces the computational complexity.

8.3.3 Distributed NMPC Implementation

For the model predictive control, it is well known that the prediction horizon affects not only the control performance but also the stability margin. Usually, we want to leave the prediction horizon a free parameter for tuning. Decreasing the number of control inputs, therefore, becomes the realistic option to alleviate the computational burden. In view of this point, the very first idea is to separate the thrust allocation (TA) from the tracking control.

Since P-5 is satisfied, the Moore–Penrose pseudoinverse method can be adopted as the closed-form solution for the TA:

$$\mathbf{u} = (\mathbf{B}^{\text{T}}\mathbf{B})^{-1}\mathbf{B}^{\text{T}}\boldsymbol{\tau} = \mathbf{B}^{+}\boldsymbol{\tau}. \tag{8.24}$$

Then we choose $\boldsymbol{\tau} = [F_u, F_v, F_r]^{\text{T}}$ as the decision variables in (P_0). As a result, the number of control inputs m reduces from 4 to 3.

To facilitate the following derivations, instead of using $|\mathbf{B}^{+}\boldsymbol{\tau}(s)| \leq \mathbf{u}_{\max}$ in place of (8.20d), we use $|\boldsymbol{\tau}(s)| \leq \boldsymbol{\tau}_{\max}$ with $\boldsymbol{\tau}_{\max} = [F_{u,\max}, F_{v,\max}, F_{r,\max}]^{\mathrm{T}}$, which is a direct bound constraint on the decision variables. Suppose Assumption 4.1 can be satisfied. Then we can take advantage of Proposition 4.1 to guarantee that the TA is always feasible for the real AUV system. According to (8.22), the expected number of flops has already dropped by approximately 1/2 of that of the original problem (8.21).

Remark 8.1 Although choosing $\tau_{\max}$ that satisfies (4.11) guarantees the TA will fit the real system, it introduces some degree of conservativeness. Using $|\boldsymbol{\tau}(s)| \leq \boldsymbol{\tau}_{\max}$ tightens the constraints in (P_0) since (4.11) is just a sufficient condition rather than the sufficient and necessary condition. However, the constraint tightening is important for the following distributed implementation because the original constraints $|\mathbf{B}^{+}\boldsymbol{\tau}(s)| \leq \mathbf{u}_{\max}$ couples the decision variables. With the tightened constraints, the decision variables $\boldsymbol{\tau} = [F_u, F_v, F_r]^{\mathrm{T}}$ are independent.

From (2.32) and (8.3) we find that when the AUV exactly tracks the reference, the dynamic equations of motion are loosely coupled. In fact, the surge dynamics (2.32a) and yaw dynamics (2.32c) are totally decoupled since $v = 0$. This observation inspires the idea of distributed implementation for solving the NMPC problem (8.21) approximately.

In the distributed control paradigm, the tracking control signals are calculated in parallel by three subsystems. The surge subsystem only considers the surge dynamics and the kinematics and determines the control input F_u by solving the following subproblem:

$$\min_{F_u \in S(\delta)} J_1 = \int_0^T (\|\tilde{\mathbf{x}}(s)\|_Q^2 + r_{11} F_u^2(s))ds + \|\tilde{\mathbf{x}}(T)\|_P^2 \tag{8.25a}$$

$$\text{s.t. } \dot{\xi}(s) = f_1(\xi(s), \hat{v}(s), \hat{r}(s), F_u(s)), \tag{8.25b}$$

$$\xi(0) = \xi(t_0), \tag{8.25c}$$

$$|F_u(s)| \leq F_{\mathrm{u},\max}, \tag{8.25d}$$

where $\xi = [x, y, \psi, u]^{\mathrm{T}}$ is the state of the surge subsystem and the subsystem model f_1 can be elaborated by taking the corresponding columns and rows in (8.19). The f_1 also contains the information of v and r but we can view them as known coefficients for some assumed trajectories $\hat{v}(s)$ and $\hat{r}(s)$.

Similarly, the sway subsystem includes the sway dynamics and the kinematics, and determines the control input F_v by solving the subproblem:

$$\min_{F_v \in S(\delta)} J_2 = \int_0^T (\|\tilde{\mathbf{x}}(s)\|_Q^2 + r_{22} F_v^2(s))ds + \|\tilde{\mathbf{x}}(T)\|_P^2 \tag{8.26a}$$

$$\text{s.t. } \dot{\zeta}(s) = f_2(\zeta(s), \hat{u}(s), \hat{r}(s), F_v(s)), \tag{8.26b}$$

$$\zeta(0) = \zeta(t_0), \tag{8.26c}$$

$$|F_v(s)| \leq F_{\text{v,max}}, \tag{8.26d}$$

where $\zeta = [x, y, \psi, v]^{\text{T}}$ is the state of the sway subsystem and the model f_2 can be obtained by taking the corresponding columns and rows in (8.19).

The yaw subsystem encompasses the yaw dynamics and the kinematics, and determines the control input F_r by solving the following optimization problem:

$$\min_{F_r \in S(\delta)} J_3 = \int_0^T (\|\tilde{\mathbf{x}}(s)\|_Q^2 + r_{33} F_r^2(s))ds + \|\tilde{\mathbf{x}}(T)\|_P^2 \tag{8.27a}$$

$$\text{s.t. } \dot{\omega}(s) = f_3(\omega(s), \hat{u}(s), \hat{v}(s), F_r(s)), \tag{8.27b}$$

$$\omega(0) = \omega(t_0), \tag{8.27c}$$

$$|F_r(s)| \leq F_{\text{r,max}}, \tag{8.27d}$$

where $\omega = [x, y, \psi, r]^{\text{T}}$ is the state of the yaw subsystem and the dynamics f_3 can be explicated by picking out the corresponding columns and rows in (8.19).

The assumed state trajectories $\hat{u}$, $\hat{v}$, and $\hat{r}$ can be determined using the following way: Let $\mathbf{F}_u^* = [F_u^*(0), \ldots, F_u^*(N-1)]^{\text{T}}$, $\mathbf{F}_v^* = [F_v^*(0), \ldots, F_v^*(N-1)]^{\text{T}}$, $\mathbf{F}_r^* = [F_r^*(0), \ldots, F_r^*(N-1)]^{\text{T}}$ be the optimal solutions of the subproblems (8.25), (8.26) and (8.27) for the previous time instant. At current time instant, construct the assumed control signals using the previous solution as

$$\hat{\mathbf{F}}_u = [F_u^*(1), \ldots, F_u^*(N-1), F_u^*(N-1)]^{\text{T}}, \tag{8.28a}$$

$$\hat{\mathbf{F}}_v = [F_v^*(1), \ldots, F_v^*(N-1), F_v^*(N-1)]^{\text{T}}, \tag{8.28b}$$

$$\hat{\mathbf{F}}_r = [F_r^*(1), \ldots, F_r^*(N-1), F_r^*(N-1)]^{\text{T}}. \tag{8.28c}$$

Then the assumed state trajectories $\hat{u}$, $\hat{v}$ and $\hat{r}$ can be obtained via state evolution through the dynamic equations (2.32) and kinematic equations (2.34) starting from the updated system state $\mathbf{x}(t)$ for the current time.

By solving the subproblems in parallel, the numerical efficiency can be improved significantly. The worst-case number of flops is down to approximately $1/30$ of that using the centralized implementation.

Remark 8.2 In general, the optimal control obtained by the above distributed implementation is not a local minimum of the original problem. To converge to a local minimum, the parallel computation needs to iterate (Bertsekas and Tsitsiklis 2015), which is obviously undesired for the additional communication and computation burden. Fortunately, for the AUV tracking control the optimum is usually not a necessity. In MPC applications, it is well known that using the warm start (8.28), the optimal solutions won't differ too much from the initial guess. Therefore, we particularly exploit this property and avoid subsystem-wide iterations during the distributed implementation.

An initialization procedure is presented to facilitate the warm start: At the very first sampling instant, solve the centralized problem (8.20) once to obtain the assumed control trajectories $\hat{\boldsymbol{\tau}}(s) = \boldsymbol{\tau}^*(s)$ for $t = 0$. The distributed control algorithm is summarized in Algorithm 8.4.

Algorithm 8.4 Distributed Implementation Algorithm

1: Initialization: Input the weighting matrices and prediction horizon; solve the NMPC problem (8.20) at t=0, let $\mathbf{u}^*(s)$ denote the solution; set $\hat{\boldsymbol{\tau}}(s) = \boldsymbol{\tau}^*(s) = \mathbf{B}\mathbf{u}^*(s)$.
2: Fetch the state measurement $\mathbf{x}(t)$.
3: Calculate the assumed state trajectories by $\dot{\hat{\mathbf{x}}} = \mathbf{f}(\hat{\mathbf{x}}, \mathbf{B}^+\hat{\boldsymbol{\tau}})$ with $\hat{\mathbf{x}}(0) = \mathbf{x}(t)$.
4: Solve the subproblems (8.25), (8.26) and (8.27) in parallel; let $\boldsymbol{\tau}^*(s) = [F_u^*(s), F_v^*(s), F_r^*(s)]^{\mathrm{T}}$ denote the solution.
5: Set $\mathbf{u}^*(s) = \mathbf{B}^+\boldsymbol{\tau}^*(s)$; construct $\hat{\boldsymbol{\tau}}(s)$ using (8.28).
6: Use $\mathbf{u}^*(s)$ for only one sampling period: $\mathbf{u}(t) = \mathbf{u}^*(s)$ for $s \in [0, \delta]$.
7: At next sampling time, set $t = t + \delta$; Goto step 2.

So far, what we focused was only the computational complexity of the algorithm, and we had no explicit consideration of the closed-loop stability. Although with enough long prediction horizon the stability can always be obtained, it is preferable to provide some analytical methods to guarantee this important closed-loop property.

The optimality of the solutions does not indicate the stability for the distributed implementation in Algorithm 8.4, not even for the centralized NMPC. To ensure it, we need to modify the subproblems and solve them sequentially.

Assume that $\mathbf{v}_d = [u_d, v_d, r_d]^{\mathrm{T}}$ obeys the dynamic equations of motion, then we can have the reference control forces $\boldsymbol{\tau}_d = [F_{ud}, F_{vd}, F_{rd}]^{\mathrm{T}}$ given by

$$\boldsymbol{\tau}_d = \mathbf{M}\mathbf{v}_d + \mathbf{C}(\mathbf{v}_d)\mathbf{v}_d + \mathbf{D}(\mathbf{v}_d)\mathbf{v}_d + \mathbf{g}(\boldsymbol{\eta}_d), \tag{8.29}$$

where $\mathbf{v}_d$ can be calculated by taking time derivative of (8.3).

Define $\tilde{F}_u = F_u - F_{ud}$, $\tilde{F}_v = F_v - F_{vd}$, and $\tilde{F}_r = F_r - F_{rd}$, and the surge subproblem is modified as follows:

$$\min_{F_u \in S(\delta)} J_1 = \int_0^T (\|\tilde{\mathbf{x}}(s)\|_Q^2 + r_{11}\tilde{F}_u^2(s))ds + \|\tilde{\mathbf{x}}(T)\|_P^2 \tag{8.30a}$$

$$\text{s.t. } \dot{\xi}(s) = f_1(\xi(s), \hat{v}(s), \hat{r}(s), F_u(s)), \tag{8.30b}$$

$$\xi(0) = \xi(t_0), \tag{8.30c}$$

$$|F_u(s)| \leq F_{\text{u,max}}, \tag{8.30d}$$

$$\tfrac{\partial V}{\partial \mathbf{x}}\mathbf{f}(\mathbf{x}(0), \lambda(0)) \leq \tfrac{\partial V}{\partial \mathbf{x}}\mathbf{f}(\mathbf{x}(0), h(\mathbf{x}(0))), \tag{8.30e}$$

where $h(\mathbf{x}) = \mathbf{B}^{+}\bar{h}(\mathbf{x})$ with $\bar{h}(\mathbf{x}) = [\bar{h}_u(\mathbf{x}), \bar{h}_v(\mathbf{x}), \bar{h}_r(\mathbf{x})]^{\text{T}}$ is an auxiliary tracking controller and $V(\mathbf{x})$ is the corresponding Lyapunov function; $\lambda(0) = \mathbf{B}^{+}\bar{\lambda}(0)$ with $\bar{\lambda}(0) = [F_u(0), \bar{h}_v(\mathbf{x}(0)), \bar{h}_r(\mathbf{x}(0))]^T$.

The modified subproblem for the sway subsystem is constructed as follows:

$$\min_{F_v \in S(\delta)} J_2 = \int_0^T (\|\tilde{\mathbf{x}}(s)\|_Q^2 + r_{22}\tilde{F}_v^2(s))ds + \|\tilde{\mathbf{x}}(T)\|_P^2 \tag{8.31a}$$

$$\text{s.t. } \dot{\zeta}(s) = f_2(\zeta(s), \hat{u}(s), \hat{r}(s), F_v(s)), \tag{8.31b}$$

$$\zeta(0) = \zeta(t_0), \tag{8.31c}$$

$$|F_v(s)| \leq F_{\text{v,max}}, \tag{8.31d}$$

$$\tfrac{\partial V}{\partial \mathbf{x}}\mathbf{f}(\mathbf{x}(0), \pi(0)) \leq \tfrac{\partial V}{\partial \mathbf{x}}\mathbf{f}(\mathbf{x}(0), h(\mathbf{x}(0))), \tag{8.31e}$$

where $\pi(0) = \mathbf{B}^{+}\bar{\pi}(0)$ and $\bar{\pi}(0) = [F_u^*(0), F_v(0), \bar{h}_r(\mathbf{x}(0))]^{\text{T}}$; $F_u^*(s)$ is the solution for (8.30) passed from the surge subsystem.

And the optimization problem for the yaw subsystem is modified as follows:

$$\min_{F_r \in S(\delta)} J_3 = \int_0^T (\|\tilde{\mathbf{x}}(s)\|_Q^2 + r_{33}\tilde{F}_r^2(s))ds + \|\tilde{\mathbf{x}}(T)\|_P^2 \tag{8.32a}$$

$$\text{s.t. } \dot{\omega}(s) = f_3(\omega(s), \hat{u}(s), \hat{v}(s), F_r(s)), \tag{8.32b}$$

$$\omega(0) = \omega(t_0), \tag{8.32c}$$

$$|F_r(s)| \leq F_{\text{r,max}}, \tag{8.32d}$$

$$\frac{\partial V}{\partial \mathbf{x}}\mathbf{f}(\mathbf{x}(0), \mathbf{u}(0)) \leq \frac{\partial V}{\partial \mathbf{x}}\mathbf{f}(\mathbf{x}(0), h(\mathbf{x}(0))), \tag{8.32e}$$

where $\mathbf{u}(0) = \mathbf{B}^{+}\boldsymbol{\tau}(0)$ and $\boldsymbol{\tau}(0) = [F_u^*(0), F_v^*(0), F_r(0)]^{\text{T}}$; $F_u^*(s)$, $F_v^*(s)$ are the solutions for (8.30) and (8.31), respectively.

The modified distributed control algorithm is summarized in Algorithm 8.5.

Algorithm 8.5 Modified Distributed Implementation Algorithm

1: Initialization: Input the weighting matrices and prediction horizon; solve the NMPC problem (8.20) at t=0, let $\mathbf{u}^*(s)$ denote the solution; set $\hat{\boldsymbol{\tau}}(s) = \boldsymbol{\tau}^*(s) = \mathbf{B}\mathbf{u}^*(s)$.
2: Fetch the state measurement $\mathbf{x}(t)$.
3: Calculate the assumed state trajectories by $\dot{\hat{\mathbf{x}}} = \mathbf{f}(\hat{\mathbf{x}}, \mathbf{B}^+\hat{\boldsymbol{\tau}})$ with $\hat{\mathbf{x}}(0) = \mathbf{x}(t)$.
4: Solve the subproblem (8.30) and send the solution $F_u^*(s)$ to the sway and yaw subsystems.
5: Solve the subproblem (8.31) and send the solution $F_v^*(s)$ to yaw subsystems.
6: Solve the subproblem (8.32) and let the solution be $F_r^*(s)$.
7: Set $\mathbf{u}^*(s) = \mathbf{B}^+\boldsymbol{\tau}^*(s)$ with $\boldsymbol{\tau}^*(s) = [F_u^*(s), F_v^*(s), F_r^*(s)]^{\mathrm{T}}$; construct $\hat{\boldsymbol{\tau}}(s)$ using (8.28).
8: Use $\mathbf{u}^*(s)$ for only one sampling period: $\mathbf{u}(t) = \mathbf{u}^*(s)$ for $s \in [0, \delta]$.
9: At next sampling time, set $t = t + \delta$; Goto step 2.

Since the subproblems are solved in sequence, the computational complexity for the modified distributed implementation is about 1/10 of that for the centralized implementation. Although it is not as efficient as Algorithm 8.4, we can establish the feasibility theorem and stability theorem for Algorithm 8.5.

Theorem 8.2 *Choosing $F_{u,max} = F_{v,max} = F_{r,max} = \tau_{max}$ which satisfies (4.11), if $\|\bar{h}(\mathbf{x})\|_\infty \le \tau_{max}$ can hold, then Algorithm 8.5 admits recursive feasibility, i.e., we can always find an initial feasible solution for each of the optimizations (8.30)–(8.32) to start with.*

Proof As $\|\bar{h}(\mathbf{x})\|_\infty \le \tau_{\max}$, we have $|\bar{h}_u(\mathbf{x})| \le F_{\mathrm{u,max}}$. Then the $\bar{h}_u(\mathbf{x})$ can be used as a feasible initial solution for the surge subproblem (8.30).

For the sway subsystem, since $|\bar{h}_v(\mathbf{x})| \le F_{\mathrm{v,max}}$ and $F_u^*(s)$ satisfies (8.30e), it can be easily verified that $\bar{h}_v(\mathbf{x})$ must be a feasible solution for the subproblem (8.31).

Likewise, we have $|\bar{h}_r(\mathbf{x})| \le F_{\mathrm{r,max}}$ and $F_u^*(s)$, $F_v^*(s)$ satisfying (8.31e). Therefore, $\bar{h}_r(\mathbf{x})$ is feasible for (8.32). □

Theorem 8.3 *Suppose that there exists an auxiliary control law $\bar{h}(\mathbf{x})$ such that $\tilde{\mathbf{x}} = \mathbf{0}$ is asymptotically stable for the closed-loop system controlled by $h(\mathbf{x}) = \mathbf{B}^+\bar{h}(\mathbf{x})$; $V(\mathbf{x})$ is the corresponding Lyapunov function. Provided that the recursive feasibility can be guaranteed, then the closed-loop system under Algorithm 8.5 is asymptotically stable and the AUV converges to the desired trajectory.*

Proof By converse Lyapunov theorems (Khalil 1996), there must be some functions $\gamma_i(\cdot)$, $i = 1, 2, 3$ which belong to class $\mathcal{K}_\infty$ such that the following inequalities hold:

$$\gamma_1(\|\mathbf{x}\|) \le V(\mathbf{x}) \le \gamma_2(\|\mathbf{x}\|), \tag{8.33a}$$

$$\frac{\partial V}{\partial \mathbf{x}}\mathbf{f}(\mathbf{x}, h(\mathbf{x})) \le -\gamma_3(\|\mathbf{x}\|). \tag{8.33b}$$

Since we have the contraction constraint (8.32e) and $\mathbf{u}(t) = \mathbf{u}^*(s)$ only for $s \in [0, \delta]$, the following condition is true:

$$\frac{\partial V}{\partial \mathbf{x}}\mathbf{f}(\mathbf{x}, \mathbf{u}(\mathbf{x})) \le \frac{\partial V}{\partial \mathbf{x}}\mathbf{f}(\mathbf{x}, h(\mathbf{x})) \le -\beta_3(|\mathbf{x}|). \tag{8.34}$$

Then by standard Lyapunov arguments ((Khalil, 1996) [Theorem 4.8]), we claim that $\tilde{\mathbf{x}} = \mathbf{0}$ is asymptotically stable for closed-loop system controlled by Algorithm 8.5. □

Theorem 8.2 and Theorem 8.3 depend on the auxiliary control law $\bar{h}(\mathbf{x})$. In principle, any Lyapunov-based tracking control law can be exploited. As an example, we can use the following backstepping control law:

$$\bar{h}(\mathbf{x}) = \mathbf{M}\dot{\mathbf{v}}_r + \mathbf{C}\mathbf{v}_r + \mathbf{D}\mathbf{v}_r + \mathbf{g} - \mathbf{R}^{\mathrm{T}}\mathbf{K}_p\tilde{\eta} - \mathbf{R}^{\mathrm{T}}\mathbf{K}_d\mathbf{s}, \tag{8.35}$$

which is derived in Section 4.3.2 and with the region of attraction (4.61).

Note that Algorithm 8.5 is not as efficient as Algorithm 8.4. However, since the closed-loop stability is always guaranteed regardless of the quality of optimal solution or the prediction horizon, we can adjust the maximum iteration number $i_{\mathrm{SQP,max}}$ (and the prediction horizon N) to achieve the real-time control. Given that the processing time t_{proc} can be fed back in real time, this adjustment can be performed automatically. A real-time DMPC algorithm is developed as Algorithm 8.6.

Algorithm 8.6 Real-Time Distributed Implementation

1: Initialization: Input the weighting matrices, prediction horizon N, maximum allowed processing time $t_{\max}$, and maximum iteration number $i_{\mathrm{SQP,max}}$; solve the NMPC problem (8.20) at t=0, let $\mathbf{u}^*(s)$ denote the solution; set $\hat{\tau}(s) = \tau^*(s) = \mathbf{B}\mathbf{u}^*(s)$.
2: Fetch the state measurement $\mathbf{x}(t)$.
3: Calculate the assumed state trajectories by $\dot{\hat{\mathbf{x}}} = \mathbf{f}(\hat{\mathbf{x}}, \mathbf{B}^+\hat{\boldsymbol{\tau}})$ with $\hat{\mathbf{x}}(0) = \mathbf{x}(t)$.
4: Solve (8.30) and send the solution $F_u^*(s)$ to the sway and yaw subsystems.
5: Solve (8.31) and send $F_v^*(s)$ to the yaw subsystem.
6: Solve (8.32) and let the solution be $F_r^*(s)$.
7: Set $\mathbf{u}^*(s) = \mathbf{B}^+\boldsymbol{\tau}^*(s)$ with $\boldsymbol{\tau}^*(s) = [F_u^*(s), F_v^*(s), F_r^*(s)]^{\mathrm{T}}$; construct $\hat{\boldsymbol{\tau}}(s)$ using (8.28).
8: Feed back the processing time t_{proc}. If $t_{\mathrm{proc}} > t_{\max}$ then set $i_{\mathrm{SQP,max}} = i_{\mathrm{SQP,max}} - \Delta i_1$; else if $t_{\mathrm{proc}} < \alpha t_{\max}$ then set $i_{\mathrm{SQP,max}} = i_{\mathrm{SQP,max}} + \Delta i_2$.
9: Implement $\mathbf{u}^*(s)$ for one sampling period: $\mathbf{u}(t) = \mathbf{u}^*(s)$ for $s \in [0, \delta]$. Set $t = t + \delta$; Goto step 2.

In Step 8 of Algorithm 8.6, we use $\alpha \in [0, 1)$, Δi_1, and Δi_2 to control the increment and decrement of the maximum iteration number. These numbers need to be well tuned. Furthermore, the automatic adjustment of the prediction horizon N can also be incorporated.

Since modern optimization algorithms (such as primal-dual interior-point methods, SQP) ensure constraint satisfaction and cost function improvement for each iteration, the Algorithm 8.6 essentially trades off between control performance and computational complexity in accordance with the current available onboard resources. Therefore, it is promising for low-cost AUVs (with low-end processors) to apply NMPC.

8.3.4 Simulation Study

In this section, we simulate the AUV to track reference trajectories to test the effectiveness of proposed distributed implementation algorithms for the NMPC tracking control method. The AUV model parameters can be found in Table 2.1. All the simulations are performed on a personal laptop (CPU: Intel(R) Core(TM) i7-3520M: 2.90GHz 2.90GHz; RAM: 4.00GB).

Two trajectories are used to test the proposed distributed implementation (DMPC). The first trajectory represents a sinusoidal path (Case I) in the horizontal plane and is defined as follows:

$$p(t) = \begin{cases} x_d = 0.5t, \\ y_d = \sin(0.5t). \end{cases} \tag{8.36}$$

The second test trajectory is an eight-shaped trajectory (Case II) defined as follows:

$$p(t) = \begin{cases} x_d = \sin(-0.5t), \\ y_d = \sin(0.25t). \end{cases} \tag{8.37}$$

For the NMPC tracking controller we use the following parameters: The sampling period $\delta = 0.1$ s; the prediction horizon $T = 5\delta$; the weighting matrices $Q = \text{diag}(10^5, 10^5, 10^3, 10^2, 10^2, 10^2)$, $R = \text{diag}(10^{-4}, 10^{-4}, 10^{-4}, 10^{-4})$, and $P = \text{diag}(10^3, 10^3, 10^2, 10, 10, 10)$; the limit on each thruster is 500 (N). The control gains $\mathbf{K}_p = \mathbf{K}_d = \text{diag}(1, 1, 1)$; and the initial condition $\mathbf{x}(0) = [0.5, 0, 0, 0, 0, 0]^{\mathrm{T}}$. For the proposed real-time distributed implementation, we set time constraint $t_{\max} = 0.1$ s, initial maximum iteration number $i_{\text{SQP,max}} = 500$, $\alpha = 0.5$, and $\Delta i_1 = \Delta i_2 = 10$.

To solve the optimization problems, we adopt the SQP method and use the embedded MATLAB function `fmincon` as the NLP solver in the simulations.

The simulation results are plotted in Figs. 8.10 and 8.11. In each plot, the blue curve is the simulated AUV trajectory using the auxiliary backstepping control (BSC); the green curve is the AUV trajectory with the centralized NMPC implementation; the magenta curve is the vehicle trajectory with the parallel distributed implementation (DMPC); the red curve is with the modified sequential distributed implementation (mDMPC); the cyan curve is the AUV trajectory generated by the real-time distributed implementation algorithm (rDMPC), and the black curve is the desired trajectory (REF). Generally speaking, similar observations can be made for the two test cases: All the tracking controllers successfully drive the vehicle converging to the desired trajectories, which demonstrates the control stability for each closed-loop system controlled by different methods. However, as mentioned several times, this important stability property is not automatically ensured by the optimality of the solution. The stability obtained for NMPC and DMPC owe to the good selection of parameters (i.e., the prediction horizon and the weighting matrices).

Figures 8.12 and 8.13 exemplify the lost stability. If the parameters are badly selected, the AUV controlled by NMPC and DMPC may not converge to the desired trajectory. As can be seen from these two figures, when we choose a bad prediction horizon, the AUV trajectories controlled by NMPC (for $N = 1$) and DMPC (for

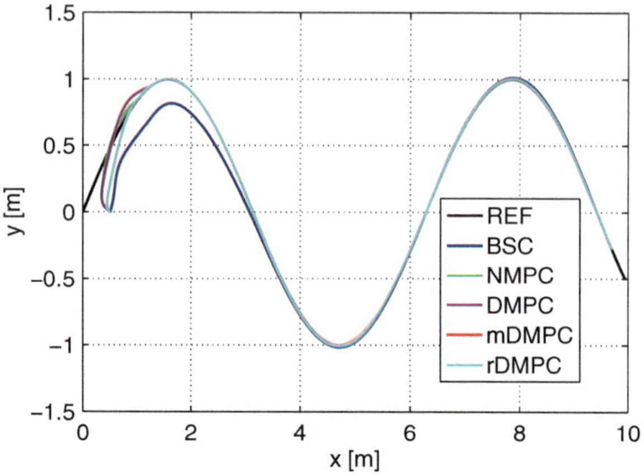

Fig. 8.10 AUV successfully tracks the desired trajectory with different control algorithms—Case I

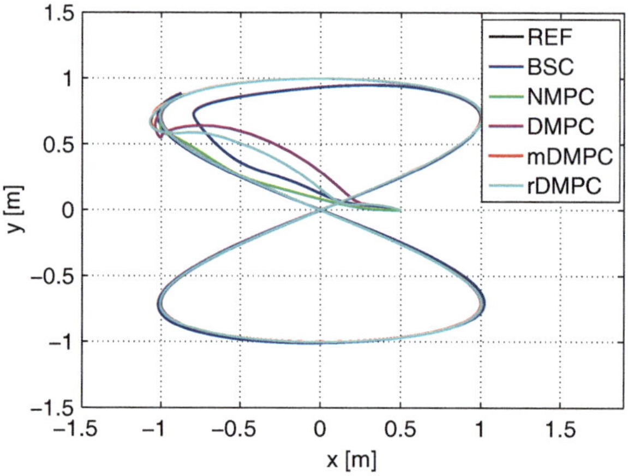

Fig. 8.11 AUV successfully tracks the desired trajectory with different control algorithms—Case II

both $N = 1$ and $N = 2$) will diverge. Similar instability can occur if the weighting matrices are badly chosen. In contrast, the mDMPC and rDMPC demonstrate the guaranteed closed-loop stability in both cases, which highlights their main advantages. Nevertheless, for tracking fixed types of desired trajectories, we are usually able to find a set of well-tuned parameters. In AUV applications that involve tracking of arbitrary trajectories (such as fully autonomous driving), it would be better to have some theoretical guarantees.

Further comparing the tracking performance (for stable cases), we find that with the MPC controllers the AUV converges faster than using the auxiliary BSC controller. This is because the MPC leverages online optimization to search for the best possible solution, while the control gains $\mathbf{K}_p$ and $\mathbf{K}_d$ for BSC are selected to be

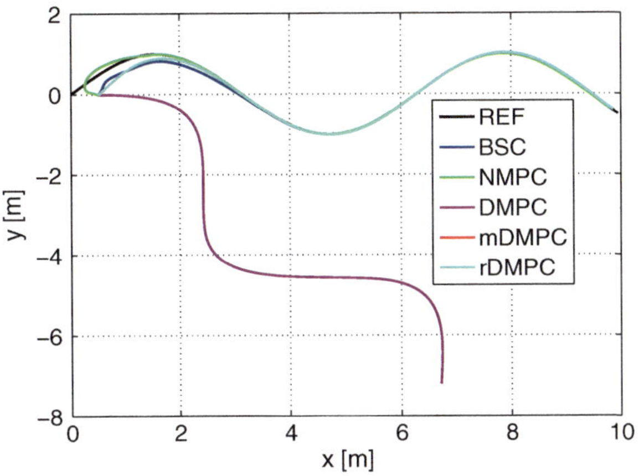

Fig. 8.12 Lost stability due to badly chosen parameters—Case I with N = 2

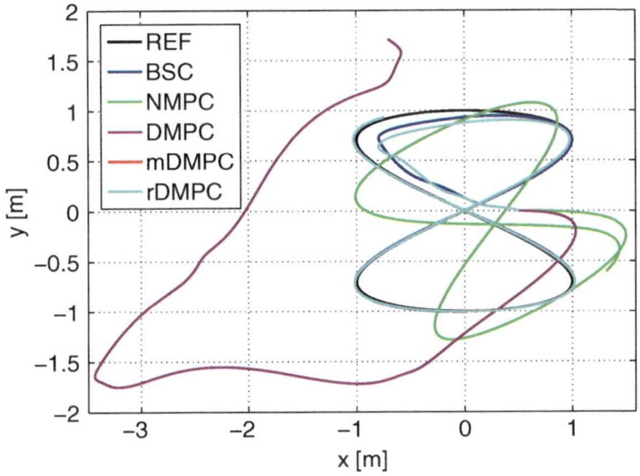

Fig. 8.13 Lost stability due to badly chosen parameters—Case II with N = 1

small for a large guaranteed region of attraction. As discussed in Remark 8.1, since we introduce certain conservativeness into the distributed implementation, the centralized NMPC should outperform the DMPC, which can be observed from Fig. 8.10 and Fig. 8.11. However, if we compare the average computing time for each implementation, as shown in Fig. 8.14, the distributed implementations demonstrate significant improvement in terms of algorithmic efficiency. The processing time for the real-time distributed implementation (rDMPC) is plotted in Fig. 8.15, which shows that the time constraint can be respected during the control.

The control commands for the thrusters are recorded in Figs. 8.16 and 8.17. We observe that at the beginning of the tracking, the MPC controllers make more aggressive moves than the BSC controller to get the fastest possible convergence. The BSC

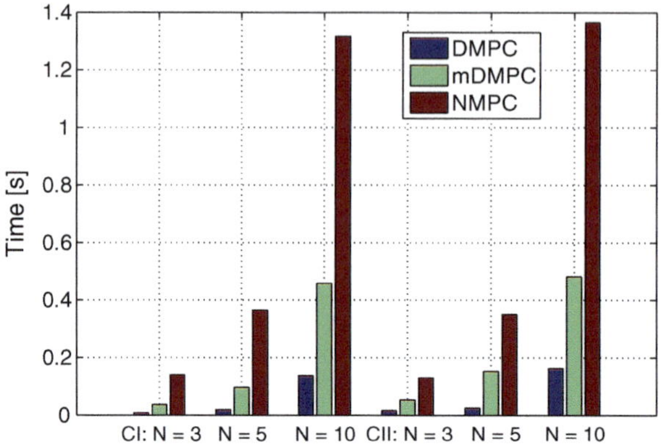

Fig. 8.14 Average computating time per update

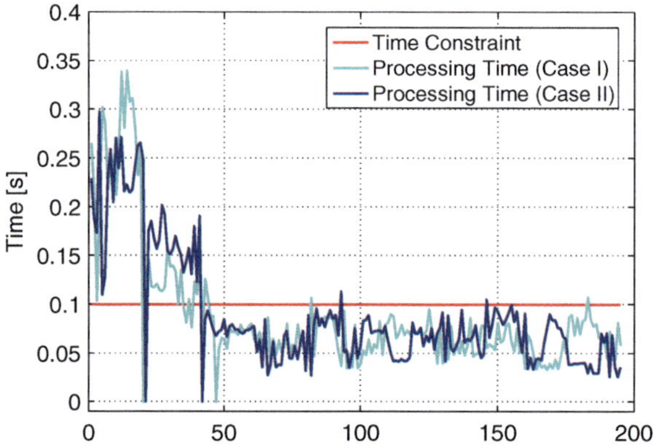

Fig. 8.15 Processing time for the proposed real-time distributed implementation

with a fixed control gain cannot fully use the onboard thrust capability, which reveals the advantage of MPC. As expected, all the control commands stay within the permitted magnitude range.

8.3.5 Robustness Test

In real-world AUV application scenarios, disturbances and uncertainties are ubiquitous. To test the robustness, we set a 20% parametric error in the obtained dynamic model of Falcon (based on which the controllers calculate the control commands) and then check the tracking performance in the presence of cyclic disturbances: $\mathbf{w} = [20\sin(t), 30\cos(0.5t), 20\sin(0.2t)]^{\mathrm{T}}$.

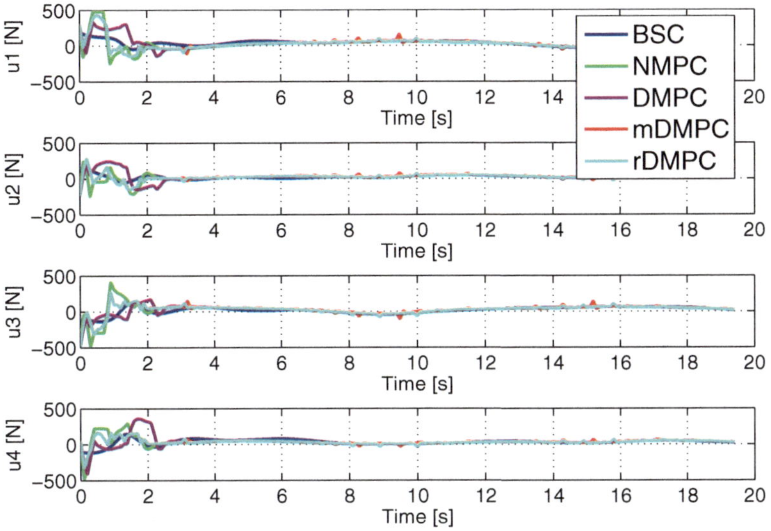

Fig. 8.16 Control command for each thruster—Case I

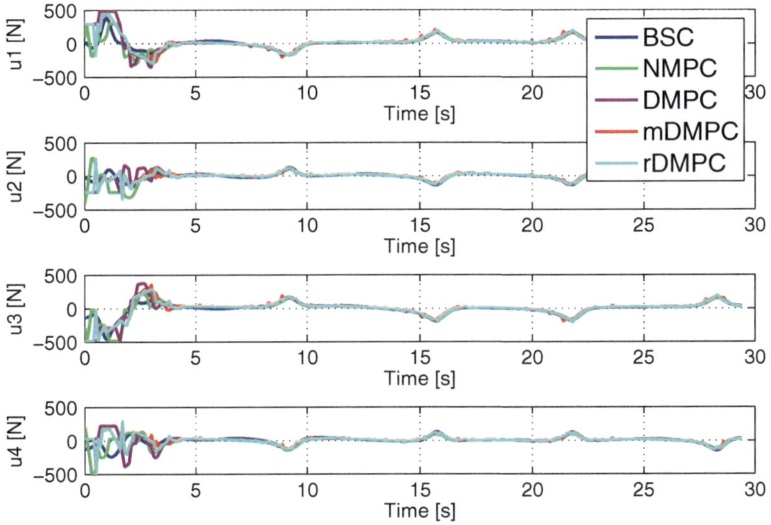

Fig. 8.17 Control command for each thruster—Case II

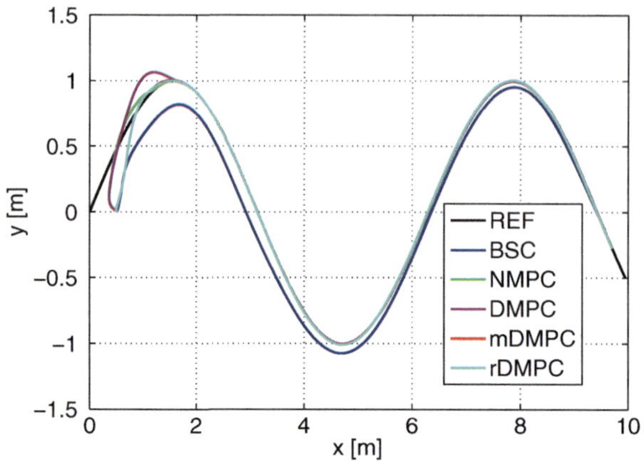

Fig. 8.18 MPC controllers are able to steer the AUV to track the desired trajectory in presence of large model errors and external disturbances—Case I

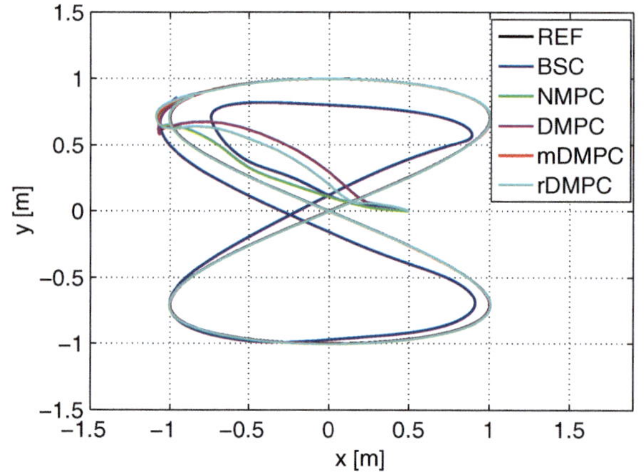

Fig. 8.19 MPC controllers are able to steer the AUV to track the desired trajectory in presence of large model errors and external disturbances—Case II

From the simulation results illustrated in Figs. 8.18 and 8.19, we observe that all MPC controllers lead the AUV converging to the desired trajectories in the presence of parametric uncertainties and external disturbances, while the BSC exhibits visible tracking error. This demonstrates the prominent robustness against parametric error and external disturbances.

For the DMPC implementation, the computation of control command for each subsystem can be performed in parallel. For each subproblem, we need to have an assumed state trajectory $\hat{\mathbf{x}}$ constructed using the solutions of neighboring subsystems (8.28). Ideally, each subsystem uses the same assumed trajectory. However, due to

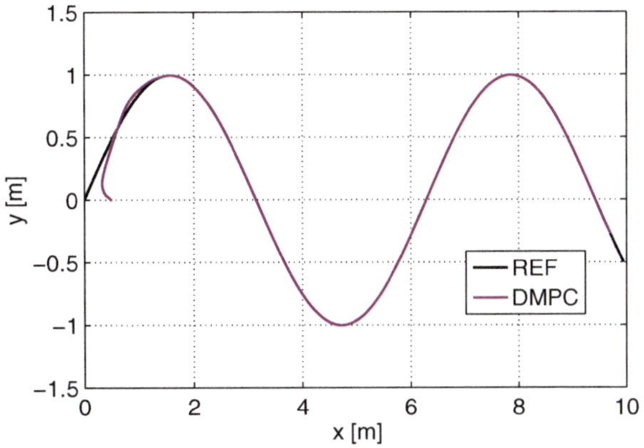

Fig. 8.20 AUV successfully tracks the desired trajectory by DMPC with mismatched assumed trajectories—Case I

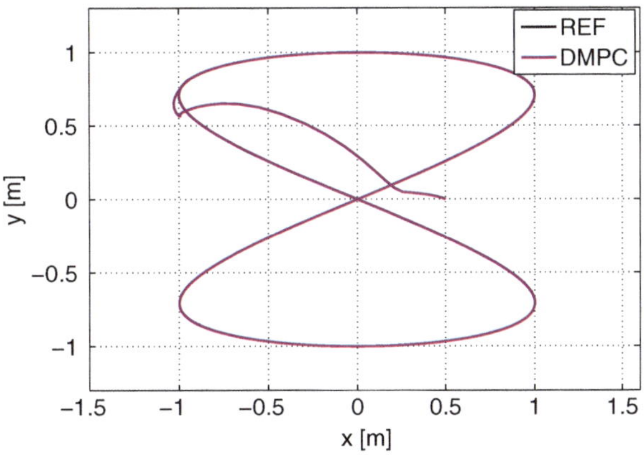

Fig. 8.21 AUV successfully tracks the desired trajectory by DMPC with mismatched assumed trajectories—Case II

communication latency, the solutions of neighboring subsystems may not be updated in time. This will result in a mismatch in the assumed trajectories, which may influence the control performance. We also test the robustness of the DMPC against such mismatch. In Figs. 8.20 and 8.21, we simulate the tracking control in the situation that each subsystem construct the assumed trajectory using the solution of its neighbors for the previous time. The simulation results demonstrate very satisfactory tracking performances, which implies excellent robustness of the proposed algorithm.

8.4 Notes and Summary

In this chapter, we have investigated the nonlinear model predictive control for the trajectory-tracking application of an autonomous underwater vehicle. To remove the computational barrier and increase the possibility of applying the NMPC to real-world AUV systems, we proposed two fast implementation strategies that dramatically reduce the computational complexity while maintaining good tracking control performance.

Firstly, the modified C/GMRES algorithm was presented. By exploiting the relationship between PMP and KKT conditions, the barrier functions were successfully incorporated into the numerical continuation approximation. The convergence of the approximate solution was proved. Extensive simulation studies revealed the effectiveness and efficiency of the proposed modified C/GMRES algorithm for the AUV trajectory-tracking control.

Secondly, a distributed implementation strategy was developed to enhance the control stability in the fast implementation of the NMPC tracking control. By exploiting the dynamic properties of the AUV motion, the optimal control signals can be determined by solving several subproblems with smaller size. Since the computational complexity increases exponentially with the problem size, the numerical efficiency can be significantly improved. A parallel implementation algorithm, a sequential implementation algorithm, and a real-time implementation algorithm were proposed. The parallel implementation algorithm (DMPC) minimizes the computational complexity, while the sequential implementation algorithm (mDMPC) sacrifices a bit of execution efficiency for theoretical guarantees of the closed-loop stability. The further developed real-time implementation algorithm (rDMPC) leverages the guaranteed closed-loop stability to satisfy the real-time control requirement by dynamically adjusting the maximum allowed computational complexity in accordance with the available computing resources. Simulation results suggest that the distributed implementation method is not only extremely efficient in executing the NMPC but also very robust against disturbances and uncertainties.

References

Allgower EL, Georg K (2012) Numerical continuation methods: an introduction, vol 13. Springer Science & Business Media

Antoniou A, Lu W-S (2007) Practical optimization. Springer, Berlin

Bertsekas D, Tsitsiklis J (2015) Parallel and distributed computation: numerical methods. Athena Scientific

Boyd S, Boyd SP, Vandenberghe L (2004) Convex optimization. Cambridge University Press

Büskens C, Maurer H (2000) SQP-methods for solving optimal control problems with control and state constraints: adjoint variables, sensitivity analysis and real-time control. J Comput Appl Math 120(1–2):85–108

Camponogara E, Jia D, Krogh BH, Talukdar S (2002) Distributed model predictive control. IEEE Control Syst Mag 22(1):44–52

Christofides PD, Scattolini R, de la Pena DM, Liu J (2013) Distributed model predictive control: a tutorial review and future research directions. Comput Chem Eng 51:21–41

Diehl M, Ferreau HJ, Haverbeke N (2009) Efficient numerical methods for nonlinear MPC and moving horizon estimation. In: Nonlinear model predictive control. Springer, Berlin, pp 391–417

Findeisen R, Allgöwer F (2004) Computational delay in nonlinear model predictive control. IFAC Proc Vols 37(1):427–432

Kelley CT (1995) Iterative methods for linear and nonlinear equations. SIAM

Khalil HK (1996) Nonlinear systems. Prentice Hall, New York

Li H, Shi Y (2014) Event-triggered robust model predictive control of continuous-time nonlinear systems. Automatica 50(5):1507–1513
McGovern LK (2000) Computational analysis of real-time convex optimization for control systems. PhD thesis, Massachusetts Institute of Technology
Mehrotra S (1992) On the implementation of a primal-dual interior point method. SIAM J Optim 2(4):575–601
Ohtsuka T (2004) A continuation/GMRES method for fast computation of nonlinear receding horizon control. Automatica 40(4):563–574
Proctor AA (2014) Semi-autonomous guidance and control of a Saab SeaEye Falcon ROV. PhD thesis
Qin SJ, Badgwell TA (1997) An overview of industrial model predictive control technology. In: AICHE symposium series, vol 93. American Institute of Chemical Engineers, New York, NY, 1971–2002, pp 232–256
Richter S, Jones CN, Morari M (2012) Computational complexity certification for real-time MPC with input constraints based on the fast gradient method. IEEE Trans Autom Control 57(6):1391–1403
Yuan Y (2000) A review of trust region algorithms for optimization. In: ICIAM, vol 99, pp 271–282

Chapter 9
Concluding Remarks

9.1 Summary of the Book

This book revolves around the MPC controller design problems for general AMV systems. Starting from the fundamental control tasks for a single AMV including the trajectory tracking, the dynamic positioning, and path-following control, we demonstrated the opportunities and challenges of using MPC to evolve next-generation marine control technology. We gradually touched onto the multi-AUV coordination and cooperation problems and proposed effective and inspiring MPC solutions. Key design considerations and trade-offs were discussed with rigorous analyses and mathematical proofs.

Chapter 3 was concerned with the combined path planning and tracking control problem for the AUV to navigate in the constrained workspace. A novel RHO framework consisting of the spline-based dynamic path planning and the NMPC trajectory-tracking control was presented. The dynamic path planner approximated a minimum curvature reference path by repeatedly solving optimization problems constructed by the current sensor information and boundary conditions descended from the previous solution, which accommodated the practically finite perceiving capability of the AUV. The planned reference path was assigned a timing law and then augmented to provide a valid reference for each state of the AUV. The error system was elaborately designed, which facilitated the stability analysis of the closed-loop tracking control systems. Sufficient conditions that guarantee the closed-loop stability were derived analytically.

In Chap. 4, the AUV dynamic positioning and the trajectory-tracking control problems were studied with the MPC setup. An LMPC framework was developed for the AUV to draw computing power (to perform online optimization) to improve the control performance. With a nonlinear PID type auxiliary control law, the LMPC-based dynamic positioning controller was elaborated. Recursive feasibility of the OCP and the stability of the closed-loop control system were analyzed. The quasi-global asymptotic stability could be claimed. The theoretical results obtained for the

Y. Shi et al., *Advanced Model Predictive Control for Autonomous Marine Vehicles*, Advances in Industrial Control, https://doi.org/10.1007/978-3-031-19354-5_9

dynamic positioning control were further extended to solve the trajectory-tracking problem. A nonlinear backstepping tracking control law was firstly derived, based on which the contraction constraint was then constructed and imposed in the OCP. The existence of the contraction constraint ensured the stability of the closed-loop LMPC tracking control system. The unique feature of the LMPC framework is that the design procedure is constructive and concise. By exploiting the existing Lyapunov-based control law, the closed-loop stability can be conveniently guaranteed. The closed-loop stability does not depend on the optimality of the solution; therefore, it essentially creates a trade-off between computational complexity and control performance for the designers to allocate computing resources.

Chapter 5 was devoted to the AUV path-following control problem. A novel MOMPC framework was proposed. Within the MOMPC framework, the different priorities between the path convergence and the speed assignment could be fully respected. Taking advantage of an explicit parametrization of the ZPE manifold, we formulated the path-following control problem into vector-valued OCPs. To solve the OCPs, two solution methods that handle objective prioritization were investigated and followed by the development of WS-MOMPC algorithm and of the LO-MOMPC algorithm, respectively. Conditions that guarantee the closed-loop stability under the two algorithms were provided, and the internal relationship between the two solution methods was explored. The proposed MOMPC method not only presents a novel framework to address the AUV path-following problem, but more importantly, it lays a theoretical foundation for the study of AUV motion control problems, which involve multiple control objectives.

Chapter 6 exemplified the application of MPC to solve multiple AUV control problems. A DMPC algorithm was developed for a multi-AUV team to perform formation-tracking mission. To handle the unknown ocean current disturbances in an explicit manner, an ESO was designed and analyzed such that the current estimation would converge to the real value. And then the current estimation could be exploited in the DMPC formation-tracking control. We showed that the DLMPC inherited the stability and robustness of the ESO-based auxiliary control law and invoked online optimization to improve formation-tracking performance of the multi-AUV system. The recursive feasibility and closed-loop stability were rigorously proved. Under the DMPC setup, the AUV only solves a local optimization problem constructed by its own state information and the neighbors'. Hence, neither the computational complexity nor the communication complexity will go unbounded with the increase in number in the AUV team.

Chapter 7 designed a robust DMPC approach for the heterogeneous ASV platoon with external disturbances over the fixed predecessor–leader following communication topology. Each follower ASV received the assumed predicted state trajectories from its predecessor and the lead ASV, sampled the local states, and broadcast the assumed predicted state trajectory to its following ASV. The local safe control input was calculated for each ASV by incorporating the coupled minimum safe following distance constraint into the DMPC optimization problem. We have rigorously proved the proposed algorithm's recursive feasibility and the ASV platoon's closed-loop stability. Simulation studies were presented to demonstrate the effectiveness of the proposed distributed platooning control method.

Different from the previous chapters which emphasize the closed-loop property of the proposed MPC solutions, in Chap. 8, an equally important ingredient of the MPC-based AUV control design: The computational complexity was investigated explicitly. Two distinct fast implementation strategies, namely the modified C/GMRES algorithm and the distributed NMPC implementation, were proposed in an attempt to further remove the potential computational barrier in applying the MPC to real-world AUV systems. The modified C/GMRES algorithm took advantage of the numerical continuation method to approximate the optimal control signals without solving the NLP problem at each sampling time instant, hence dramatically reducing the computing time. The barrier function was proposed to handle the inequality constraints in terms of thrust limits. Under differentiability and LICQ, the validity of the incorporation of the barrier functions into the numerical continuation approximations could be shown. To further preserve the closed-loop stability in the fast implementation, in this chapter, another strategy named distributed NMPC implementation was also proposed for the AUV trajectory-tracking control. The distributed implementation exploited the dynamic properties of the AUV motion and calculated one control command at a time by solving well-defined subproblems with smaller size. Since the computational complexity increases exponentially with the problem size, the proposed distributed implementation method can be numerically efficient. By imposing additional constraints in these subproblems and by solving them sequentially, the stability of the closed-loop system under the distributed NMPC implementation could be guaranteed. The work in Chap. 8 reveals the possibility of eliminating the computational cost of implementing NMPC while guaranteeing the closed-loop stability theoretically. Thus, it is expected that more novel and efficient MPC solutions will be inspired for various AUV applications.

9.2 Future Missions

The work presented in this book focuses on the MPC controller design to solve the AMV motion control problems and the design of efficient numerical algorithms to implement these MPC controllers. The obtained results attempt to provide some deep insights into the application of MPC to the AMV systems. The theoretical findings have been verified through extensive simulations, and their empirical validation can be further carried out to showcase the strengths of this promising marine control technology. Nevertheless, the MPC solution is relatively new to the marine vehicle systems, and there are many interesting yet challenging open problems deserving further studies.

9.2.1 Some Fundamental Questions

- **On the system model.** MPC utilizes the system model to make predictions and optimize the predicted system trajectory to calculate the control signal. Therefore, the model is essential. There are different ways of modeling. A rough anatomy divides the modeling methods into the first-principle approach and data-driven approaches. Conventional AUV modeling uses the first-principle approach and derives deterministic models applying laws of physics (as described in Chap. 2). The bright side of this approach may go to the availability of theoretical analysis of closed-loop performances following the mainstream design procedures in general MPC theory. The downside lies in the need of significant effort to perform parameter identification with a number of carefully designed experiments. The recent success in the field of machine learning (e.g., deep learning) has led to a growing interest in using data-driven approaches to perform modeling and control of the system. As discussed in Hewing et al. (2020), this conceptually distinct treatment in MPC context leverages the MPC formulation to solely handle constraints satisfaction, while the closed-loop performances are optimized using data-driven techniques, such as reinforcement learning algorithms. Therefore, it is extremely interesting to study the MPC for AUV systems using different modeling approaches. Furthermore, the marine environment (e.g., the current flows, waves and winds) can also be incorporated and handled in the MPC framework. Many interesting research problems can be explored in this direction.
- **On the robustness of the MPC control.** Uncertainties are ubiquitous. During the control process, uncertainties may arise in different ways. As we discussed in previous chapters, model mismatches, external disturbances, and measurement errors are uncertainties sources in practical AUV application scenarios. Fortunately, we have found, through extensive simulations, that the MPC controllers have certain degrees of inherent robustness. Yet, it is still highly desired to study the robustness of MPC controllers and modify the nominal MPC control design by explicitly taking into account the uncertainties. Robust MPC (Bemporad and Morari 1999) and stochastic MPC (Mesbah 2016) present powerful design tools to deal with the discrepancy between the predicted behavior and actual behavior of the system due to uncertainties. For AUV control applications, additional uncertainties from unexpected solver failures, from the availability of sensing information, and from communication delays to be studied in a systematic way.
- **On the effectiveness of the prediction.** Another interesting problem in MPC control design is how long the prediction horizon (and control horizon) we should use. In most cases, the prediction horizon is an important tuning parameter. Generally speaking, using longer prediction horizons gains better closed-loop performance. At the same time, however, the computational complexity grows exponentially as the prediction horizon increases (as we have discussed in Chap. 8). In the process control (which is the most mature MPC application field), a rule of thumb is $N\delta = t_s$ where δ is the sampling period, and t_s is the settling time for the open-loop response (Seborg et al. 2016). Since most industrial chemical process dynamics are

slow, the corresponding N is between 30 and 120. For fast dynamic systems, such as robotic systems, the choice of prediction horizon can be tricky. The authors in Nguyen et al. (2021) report two interesting observations: (i) For prediction horizon longer than a threshold value (2 s for UAVs), the increase of prediction horizon will not improve the control performance but only introduce additional complexity; (ii) the nonlinear MPC outperforms linear MPC consistently. Therefore, it is interesting to study the practical design considerations along this research line, and to answer the questions like: How do we characterize the effectiveness of the prediction in MPC particularly for AUV systems? More generally, providing a certain degree of accuracy on the system model, how do we make the trade-offs between complexity and performance?

- **On practical sampling, sensing, and control strategies.** In MPC, the continuous system dynamics appeared in the optimization formulation must be discretized, and then the optimization problem is solved numerically using iterative algorithms. The discretization and sampling introduce errors in the state prediction, hence influencing the closed-loop performance in a negative way. In practical MPC design, often, the effect of discretization and sampling is neglected, and the choice of the sampling period is made after the controller design through extensive experiments (Wysocki and Lawrynczuk 2014). It is possible, however, to characterize the relationship between the discretization accuracy and the performance bound using a sampled-data system control formulation (Worthmann et al. 2014). For AUV systems, the sampling rate is capped by the state-of-the-art sensing technology (e.g., the update rate of a USBL positioning system is no more 2 Hz). Therefore, the integration of state estimation (and periodically correction) into the MPC design needs to be further studied. On the other hand, if the MPC is implemented on resource-constrained embedded computer systems (such as low-cost AUVs), the sampling rate and actuation rate may need to be reduced purposely. The ideas of event-triggered and self-triggered control (Heemels et al. 2012) can be applied to the MPC controller design.
- **On modular and hierarchical designs.** Modular and hierarchical designs are favorable in engineering practice. Modular designs are commonly seen in large-scale industrial systems, where the manipulation variables (control u) and controlled variables (output y) can be grouped into nonoverlapping (or small overlapping) subsets. When such a grouping is possible, the modular design of the overall control system can be done following the decentralized/distributed control scheme. Local controllers/regulators are design separately for each group (subsystem). In the context of MPC, the advantage of modular design is obvious (Scattolini 2009): Both design complexity and run-time computational complexity can be reduced significantly. The distributed implementation strategy in Chap. 8 shows the potential to apply modular designs to MPC-based marine vehicle control. Likewise, the hierarchical design aims to improve the design tractability, and to simplify the controller synthesis and analysis. In hierarchical control, the output of upstream controller provides set-points of the downstream controller. The RHO path planner and NMPC controller in Chap. 3 can be viewed as an example of this control structure. The upstream controller (i.e., the path planner) ignores the AUV dynamics

(or only treats it indirectly by minimizing the path curvature) in order to reduce the complexity. For the motion control, the hierarchical control paradigm can also be applied. MPC can be designed at the velocity-level, and the force-level control can be fulfilled by conventional PID controllers. Thus, it is interesting to design, validate, and compare modular, hierarchical MPC controllers and their combinations for AMV systems.

9.2.2 Some Tractable AMV Research Directions

For the readers who are interested in diving into the ocean of MPC-based AMV control, we list the following tractable research topics to start off.

- **MPC-based AMV control design with incremental input constraints.** All the MPC controllers designed in this book only consider the bound input constraint which represents the thrust limits on the real AMV system. However, this is a rather simplified problem formulation. As we can see from Fig. 4.3, for the LMPC-based DP control, the control signal varies too rapidly, which is undesired in real control applications. The fast varying control command is either forbidden by the real thruster dynamics or wearing out the actuators. In practice, the change rate of the control input should be considered as well. As shown in Yu et al. (2017), despite the assumptions, the incremental input constraints can also be handled by the NMPC technique. Therefore, it is of great interest to extend the results obtained in this book to incorporate this realistic type of control inputs.
- **6-DOF AUV control design using LMPC.** The control problems addressed in this book are concerned with the AUV motion in the local level plane. This is partially due to the physical limitation of the experimental platform Falcon and partially due to the complexity of the motion control problem itself. When considering the full 6-DOF dynamic model, the couplings among the DOFs may be too complicated for the standard NMPC to handle. However, as demonstrated in Chap. 4, with the LMPC framework the closed-loop stability can be conveniently guaranteed with the help of an auxiliary Lyapunov-based control law. As there are many existing results can be exploited, it is quite interesting to explore the LMPC-based control structure, possibly with distributed implementation algorithms, for the AUV motion control problem in the general 6-DOF settings.
- **DMPC for multi-AMV cooperative and coordination control with delay tolerance.** The cooperative control of multi-agent systems has been a focused research theme for decades. DMPC presents an effective and promising tool to design the coordination control laws among the agents (Mayne 2014). The formation control in Chap. 6, the platooning control in Chap. 7, and the distributed NMPC implementation algorithms in Chap. 8 have demonstrated the power of DMPC. However, the communication delay is a practical issue to be addressed for underwater coordination control applications. In the distributed NMPC implementation algorithm (Sect. 8.3), we see the inherent robustness of the NMPC against delays in updat-

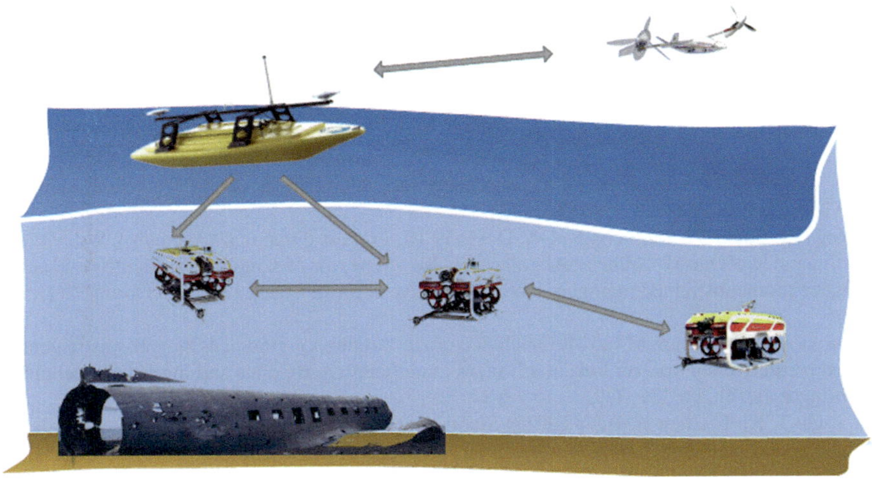

Fig. 9.1 Heterogeneous vehicles cooperate to perform search and rescue applications

ing the neighbor AUVs' states. Revealing this robustness margin and proposing strategies to enlarge the margin would be interesting from both theoretical and practical points of view.

- **Distributed MPC for heterogeneous multi-vehicle systems.** To move forward, it would be extremely interesting to explore the DMPC for cooperative and coordination control among heterogeneous multi-vehicle systems. In Chap. 7, the one-dimensional platooning control for a group of heterogeneous ASVs has been studied and solved with a robust DMPC formulation. The same idea can be extended to the general two-dimensional and three-dimensional motion control with or without the distributed implementation strategy proposed in Chap. 8. Furthermore, the stochastic DMPC (Mesbah 2016) also presents a suitable framework for solving the multi-vehicle cooperative control problems in the unpredictable marine environment. Within the powerful DMPC framework, many meaningful marine vehicle applications, such as the search and rescue task illustrated in Fig. 9.1, can be performed intelligently and autonomously.

References

Bemporad A, Morari M (1999) Robust model predictive control: a survey. In: Robustness in identification and control. Springer, Berlin, pp 207–226

Heemels WPMH, Johansson KH, Tabuada P (2012) An introduction to event-triggered and self-triggered control. In: Proceedings of 51st IEEE conference on decision and control (CDC). IEEE, pp 3270–3285

Hewing L, Wabersich KP, Menner M, Zeilinger MN (2020) Learning-based model predictive control: toward safe learning in control. Ann Rev Control Rob Auton Syst 3:269–296

Mayne DQ (2014) Model predictive control: recent developments and future promise. Automatica 50(12):2967–2986
Mesbah Ali (2016) Stochastic model predictive control: an overview and perspectives for future research. IEEE Control Syst Mag 36(6):30–44
Nguyen H, Kamel M, Alexis K, Siegwart R (2021) Model predictive control for micro aerial vehicles: a survey. In: 2021 European control conference (ECC). IEEE, pp 1556–1563
Scattolini R (2009) Architectures for distributed and hierarchical model predictive control—a review. J Process Control 19(5):723–731
Seborg DE, Edgar TF, Mellichamp DA, Doyle FJ III. Process dynamics and control. Wiley (2016)
Worthmann K, Reble M, Grüne L, Allgoöwer F (2014) The role of sampling for stability and performance in unconstrained nonlinear model predictive control. SIAM J Control Optim 52(1):581–605
Wysocki A, Ławryńczuk M (2014) On choice of the sampling period and the horizons in generalized predictive control. In: Recent advances in automation, robotics and measuring techniques. Springer, Berlin, pp 329–339
Yu S, Qu T, Xu F, Chen H, Hu Y (2017) Stability of finite horizon model predictive control with incremental input constraints. Automatica 79:265–272

Appendix A

A.1 Forward Difference Generalized Minimal Residual Method

In Sect. 8.2, the forward difference generalized minimal residual (FDGMRES) method is applied in Algorithm 8.2 in Chap. 8 to numerically solve the linear equation $F_U \dot{U} = A_s F - F_x \dot{x} - F_t$ which is approximated by the following equation:

$$D_\delta F(U, x + \delta \dot{x}, t + \delta : \dot{U}, 0, 0) = b(U, x, \dot{x}, t), \tag{A.1}$$

where

$$b(U, x, \dot{x}, t) = A_s F(U, x, t) - D_\delta F(U, x, t : 0, \dot{x}, 1). \tag{A.2}$$

The detailed FDGMRES is summarized in Algorithm A.1. It is worth noting that the generalized minimal residual method (GMRES) is a kind of Krylov subspace method to iteratively solve linear equation systems in forms of $Ax = b$. At kth iteration, GMRES searches for $x \in x_0 + \mathcal{K}_k$ such that the residual $\rho = \| b - Ax \|$ is minimized, where $\mathcal{K}_k = \text{span}\{r_0, Ar_0, \ldots, A^{k-1} r_0\}$ with $r_0 = b - Ax_0$. To facilitate the residual minimization, Gram–Schmidt process (STEP 5–STEP 11 in Algorithm A.1) is performed, so that the original minimization in Krylov subspace is transformed into an unconstrained least squares problem (STEP 13 in Algorithm A.1). The least squares problem can be solved efficiently with QR decomposition or singular value decomposition. One important feature of GMRES is that for a large linear equation system, usually a small $k_{\max}$ is enough to give a satisfactory approximate solution with the specified error tolerance.

Y. Shi et al., *Advanced Model Predictive Control for Autonomous Marine Vehicles*, Advances in Industrial Control, https://doi.org/10.1007/978-3-031-19354-5

Algorithm A.1 $\dot{U} = \text{FDGMRES}(U, x, \dot{x}, t, \hat{\dot{U}}, \delta, k_{\max})$

$\hat{r} = b(U, x, \dot{x}, t) - D_\delta F(U, x + \delta\dot{x}, t + \delta : \hat{\dot{U}}, 0, 0)$.
$v_1 = \hat{r} / \| \hat{r} \|$, $\rho = \| \hat{r} \|$, $\beta = \rho$, $k = 0$.
for $k < k_{\max}$ **do**
  $k = k + 1$
  $v_{k+1} = D_\delta F(U, x + \delta\dot{x}, t + \delta : v_k, 0, 0)$
  for $j = 1, \ldots, k$ **do**
    $h_{jk} = v_{k+1}^{\mathrm{T}} v_j$
    $v_{k+1} = v_{k+1} - h_{jk} v_j$
  end for
  $h_{k+1,k} = \| v_{k+1} \|$
  $v_{k+1} = v_{k+1} / \| v_{k+1} \|$
  $e_1 = [1, 0, \ldots, 0]^{\mathrm{T}} \in \mathbb{R}^{k+1}$, $H_k = \{h_{ij}\} \in \mathbb{R}^{(k+1)\times k}$ ($h_{ij} = 0$ for $i > j + 1$)
  Minimize $\| \beta e_1 - H_k y^k \|$ to determine $y^k \in \mathbb{R}^k$
  $\rho = \| \beta e_1 - H_k y^k \|$.
end for
$\dot{U} = \hat{\dot{U}} + V_k y^k$, where $V_k = [v_1, \ldots, v_k] \in \mathbb{R}^{mN \times k}$.

A.2 Derivation of Jacobians

In Sect. 8.2, the Jacobian matrices need to be calculated for the execution of mC/GMRES algorithm. Specifically, for the NMPC trajectory-tracking control problem (8.4), we have inequality constraints $|F_u| \le F_{u,\max}$, $|F_v| \le F_{v,\max}$, and $|F_r| \le F_{r,\max}$, which are equivalent to

$$F_u^2 - F_{u,\max}^2 \le 0 \tag{A.3a}$$
$$F_v^2 - F_{v,\max}^2 \le 0 \tag{A.3b}$$
$$F_r^2 - F_{r,\max}^2 \le 0. \tag{A.3c}$$

Hence we have the modified Hamiltonian

$$\begin{aligned} H = {} & q_{11}(x - x_d)^2 + q_{22}(y - y_d)^2 + q_{33}(\psi - \psi_d)^2 + q_{44}(u - u_d)^2 + q_{55}(v - v_d)^2 \\ & + q_{66}(r - r_d)^2 + r_{11} F_u^2 + r_{22} F_v^2 + r_{33} F_r^2 - \gamma_k \log(F_{u,\max}^2 - F_u^2) - \gamma_k \log(F_{v,\max}^2 \\ & - F_v^2) - \gamma_k \log(F_{r,\max}^2 - F_r^2) + \bar{\lambda}_1 u \cos\psi - \bar{\lambda}_1 v \sin\psi + \bar{\lambda}_2 u \sin\psi + \bar{\lambda}_2 v \cos\psi \\ & + \bar{\lambda}_3 r + \bar{\lambda}_4 \frac{F_u}{M_{\dot{u}}} + \bar{\lambda}_4 \frac{M_{\dot{v}}}{M_{\dot{u}}} vr - \bar{\lambda}_4 \frac{X_u}{M_{\dot{u}}} u - \bar{\lambda}_4 \frac{D_u}{M_{\dot{u}}} |u| u + \bar{\lambda}_5 \frac{F_v}{M_{\dot{v}}} - \bar{\lambda}_5 \frac{M_{\dot{u}}}{M_{\dot{v}}} ur \\ & - \bar{\lambda}_5 \frac{Y_v}{M_{\dot{v}}} v - \bar{\lambda}_5 \frac{D_v}{M_{\dot{v}}} |v| v + \bar{\lambda}_6 \frac{F_r}{M_{\dot{r}}} + \bar{\lambda}_6 \frac{M_{\dot{u}} - M_{\dot{v}}}{M_{\dot{r}}} uv - \bar{\lambda}_6 \frac{N_r}{M_{\dot{r}}} r - \bar{\lambda}_6 \frac{D_r}{M_{\dot{r}}} |r| r \end{aligned}$$

Then we calculate the Jacobians $H_{\mathbf{x}}$ and $H_{\mathbf{u}}$ in the following:

$$H_{\mathbf{x}} = \begin{bmatrix} 2q_{11}(x - x_d) \\ 2q_{22}(y - y_d) \\ 2q_{33}(\psi - \psi_d) - \bar{\lambda}_1 u \sin\psi - \bar{\lambda}_1 v \cos\psi + \bar{\lambda}_2 u \cos\psi - \bar{\lambda}_2 v \sin\psi \\ 2q_{44}(u - u_d) + \bar{\lambda}_1 \cos\psi + \bar{\lambda}_2 \sin\psi - \bar{\lambda}_4 \frac{X_u}{M_{\dot{u}}} - \bar{\lambda}_5 \frac{M_{\dot{u}}}{M_{\dot{v}}} r + \bar{\lambda}_6 \frac{M_{\dot{u}} - M_{\dot{v}}}{M_{\dot{r}}} v - 2\bar{\lambda}_4 \frac{D_u}{M_{\dot{u}}} |u| \\ 2q_{55}(v - v_d) - \bar{\lambda}_1 \sin\psi + \bar{\lambda}_2 \cos\psi + \bar{\lambda}_4 \frac{M_{\dot{v}}}{M_{\dot{u}}} r - \bar{\lambda}_5 \frac{Y_v}{M_{\dot{v}}} + \bar{\lambda}_6 \frac{M_{\dot{u}} - M_{\dot{v}}}{M_{\dot{r}}} u - 2\bar{\lambda}_5 \frac{D_v}{M_{\dot{v}}} |v| \\ 2q_{66}(r - r_d) + \bar{\lambda}_3 + \bar{\lambda}_4 \frac{M_{\dot{v}}}{M_{\dot{u}}} v - \bar{\lambda}_5 \frac{M_{\dot{u}}}{M_{\dot{v}}} u - \bar{\lambda}_6 \frac{N_r}{M_{\dot{r}}} - 2\bar{\lambda}_6 \frac{D_r}{M_{\dot{r}}} |r| \end{bmatrix}$$

$$H_{\mathbf{u}} = \begin{bmatrix} 2r_{11}F_u + \frac{\bar{\lambda}_4}{M_{\dot{u}}} + \gamma_k \frac{2F_u}{F_{u,\max}^2 - F_u^2} \\ 2r_{22}F_v + \frac{\bar{\lambda}_5}{M_{\dot{v}}} + \gamma_k \frac{2F_v}{F_{v,\max}^2 - F_v^2} \\ 2r_{33}F_r + \frac{\bar{\lambda}_6}{M_{\dot{r}}} + \gamma_k \frac{2F_r}{F_{r,\max}^2 - F_r^2} \end{bmatrix}.$$

Also we have $g_{\mathbf{x}} = [2q_{f11}(x - x_d), 2q_{f22}(y - y_d), 2q_{f33}(\psi - \psi_d), 2q_{f44}(u - u_d), 2q_{f55}(v - v_d), 2q_{f66}(r - r_d)]^{\mathrm{T}}$, where q_{ii}, r_{ii}, and q_{fii} are the corresponding diagonal elements of weighting matrices Q, R, and Q_f.

A.3 Introduction to MPC Simulations

All the theoretical results obtained in this book are validated through computer simulations. The simulations are based on the scientific computing software MATLAB® which can be obtained from:

The MathWorks Inc.
https://www.mathworks.com/

All the simulation codes are free of charge and can be downloaded upon request from:

Applied Control and Information Processing Lab
https://www.uvic.ca/research/labs/acipl/publications/index.php

The m-code is primarily developed on a Linux machine using MATLAB 2020a. The code should work on any platform such as Windows and MacOS, but is developed and thus most extensively tested on Linux. In order to achieve better code reusability, and to make the code structurally clear and easier to be understood, the simulation examples are coded following the Object-Oriented Programming (OOP) paradigm. Therefore, common classes and library functions (such as AUV, point2D, MPC_controller, RefGen, etc.) can be shared between chapters. The OOP coding style makes the code modular, and the readers can test and compare performances by simply replacing the controller module with different controllers. Meanwhile, to make the simulation codes extendable, several abstract classes (such as Model, Controller, Trajectory2D, etc.) are created.

Interested readers will find it convenient to extend the simulation scenarios by writing customized modules which implement the abstract classes, and then plug and play.

All classes and library functions are well-annotated. Readers can find clear explanations and instructions by peeking the source code or typing

```
>> help function-name
```

in the MATLAB command window. The simulation codes will be maintained and updated on a frequent basis as indicated by the book version number.

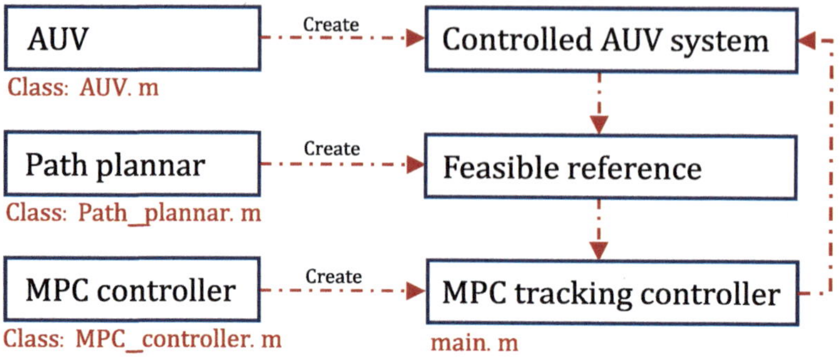

Fig. A.1 Illustration of the software implementation

A.3.1 Illustration of the Software Implementation

Here we provide a detailed description for the software implementation of the simulation example in Chap. 3. As shown in Fig. A.1, a straightforward implementation of Algorithm 3.1 in Chap. 3 is provided and illustrated for the integrated path planning and tracking control of the AUV in this subsection. We provide several function classes, such as AUV and MPC_controller to increase the reusability of the code, which will be elaborated in the sequel.

A.3.1.1 Creating a Controlled AUV System

In this subsection, we demonstrate how to set up the `AUV` class for the given example in the simulation section. To build up the `AUV` model in the simulation, the following variables pertinent to the model, including the model parameters, the number of degree, the initial state, and the initial control input, are named as `coef`, `ndof`, `X0`, and `U0`, respectively. Upon call of

```
auv = AUV(coef, ndof, X0, U0); % Create a controlled
AUV system
```

the class generates a controlled AUV object. Note that the 3-DOF AUV system model considered in Chap. 3 is easily extended to the more complex case, such as the AUV system with 6-DOF dynamics. One can create a controlled AUV by specifying system parameters when calling the `AUV` object, for example, the corresponding model parameters of the AUV in this chapter are chosen as `coef = [116.0; 13.1;-167.6;-477.2;-15.9;26.9;35.8;3.5;241.3;503.8;76.9]`.

The `AUV` class provides different functions for the controlled AUV system, which mainly consists of the following parts:

(1) Update the nominal system state without considering disturbances

```
X' = auv.dynamics_discrete(X, U, dt);
```

This function represents the nominal system model that is used in the MPC controller design for prediction and optimization. Here `dt` is the sampling interval, `X` is the measurement of the system state, and `U` is the control input computed by solving Problem 3.3.

(2) Update the actual system state

```
X = auv.advance(U, W, dt);
```

In the AUV model that has been created, we use `W` to represent the practical disturbances in the actual system.

(3) Output the error system state

```
Xe = auv.ErrorState(X, Xr);
```

Here, the `ErrorState` function transforms the original AUV dynamics into the error dynamics of AUV. Through such a transformation, the tracking control problem of the AUV is transformed into a regulation problem. `Xr` is the feasible reference trajectory generated by the path planner.

(4) Output the reference control input

```
[f1, f2, f3] = auv.F(X, Xe, P);
```

Finally, the reference control inputs are calculated as in (3.24).

A.3.1.2 Creating an MPC Controller

An `MPC_controller` following arguments: the prediction horizon, the AUV object, the weighting matrices, the upper input bound and the lower input bound; they are named as `N`, `model`, `weights`, `Umax0`, and `Umin0`, respectively. Here, an MPC object can be conveniently created by specifying these input arguments when calling `MPC_controller`:

```
mpc = MPC_controller(N, auv, weights, Umax0, Umin0); %
Create an MPC controller for the AUV
```

In the class of `MPC_controller`, we describe how to formulate an MPC optimization problem, including the objective function design, the constraint construction, and the solver usage.

(1) Objective function
The objective function in (3.15), consisting of the weighting matrices `P`, the initial states `X0`, the sampling time `dt`, and the control input sequence `u`, is designed and presented as follows:

```
cost = mpc.mpc_cost(u, X0, P, dt);
```

Note that part of the objective function is also the penalty for the control inputs. In addition, one can smoothen the obtained optimal control inputs by considering the incremental change of the control input $\Delta u(k) = u(k) - u(k-1)$. A quadratic penalty on the change can be considered in the objective function design. Alternatively, a constraint of the control input change can be designed and incorporated into the MPC optimization problem.

(2) Constraints
MPC has the advantage of handling both state and control input constraints. In particular, we consider the control input constraint and the terminal constraint for the MPC optimization problem in this book. The control input constraints are provided to the solver `fmincon` by designing the upper bound and lower bound as

```
u_max = mpc.calc_upperbound(upperbound);
```

and

```
u_min = mpc.calc_lowerbound(lowerbound);
```

Note that the output of the above two functions are the bounds of the control input sequence. Since MATLAB's optimization routine `fmincon` distinguishes the equality and inequality. Here we add inequality terminal constraints (3.19) in the MPC optimization problem.

```
[c, ceq] = mpc.mpc_constraints(u, X0, P, dt);
```

Note that `c` represents the inequality constraint for the terminal set, and the equality constraint `ceq` is empty, i.e., the equality constraints are not considered in this MPC optimization problem.

(3) Generate the optimal control input
Calling the following function, the optimal control input can be generated and will be applied to control the AUV.

```
u=mpc.calc_control(u0, X0, P, dt);
```

Together with the initial feasible control input u0, the initial measured AUV state X0, the reference trajectory P, and the sampling interval dt are supplied as arguments in the function calc_control. Note that fmincon is used as the solver to solve the MPC optimization problem; more details can be found in the MATLAB help for fmincon.

A.3.1.3 Creating a Planner Object

The first step of configuring the AUV planner is to call the Path_planner class by passing a list of input arguments, including the prediction horizon (N), the AUV object (model), and the order (mlti).

```
planner = Path_planner(N, auv, mlti); % Create a path
planner for the AUV
```

The following Table A.1 provides a list of the ingredients of the optimization problem for Path_planner and the corresponding functions in the MATLAB implementation.

Note that the superscript symbol $*$ indicates the condition for the spline function at the start point x_0 in the above table, and bbrks denotes the number of the sampling points.

The Path_planner class for the controlled AUV provides functions that mainly consists of the following components:

(1) Generate the planned path point at initial time

```
x = planner.calc_path1(X0, bbrks, X, Y, dt);
```

(2) Generate the planned path point

```
x = planner.calc_path2(x0, bbrks, X, Y, y0, y0d,
y02d, eps, dt);
```

Table A.1 Ingredients of the MPC-based planning problem

Ingredients	Function in the implementation
$F(\alpha, x)$	planning_cost(a, bbrks, dt)
$\|S^*(\alpha, x_0) - S_0(x_0)\| \leq \epsilon_1$	planning_init_constraints(a, bbrks, dt, X, Y)
$c_1(x) \leq S(\alpha, x) \leq c_2(x)$	planning_constraints(a, bbrks, dt, X, Y, y0, y0d, y02d, eps)

(3) Output the planned reference path for the AUV

```
[eta_Ref, P] = planner.reference(t, Spline_map2,
SpAry, i, M, T_plt, y_scale, dt);
```

For the simulation of the examples in this book, we provide all related M-files on our webpage https://github.com/ACIPL/MPC4AMVs.

Index

Y. Shi et al., *Advanced Model Predictive Control for Autonomous Marine Vehicles*, Advances in Industrial Control, https://doi.org/10.1007/978-3-031-19354-5

GPSR Compliance

The European Union's (EU) General Product Safety Regulation (GPSR) is a set of rules that requires consumer products to be safe and our obligations to ensure this.

If you have any concerns about our products, you can contact us on ProductSafety@springernature.com

In case Publisher is established outside the EU, the EU authorized representative is:

Springer Nature Customer Service Center GmbH
Europaplatz 3
69115 Heidelberg, Germany

Batch number: 10726135

Printed by Printforce, the Netherlands